WITHDRAWN

509.17671 I64
Iqbal, Muzaffar
Science and islam
67181

Science and Islam

**Recent Titles in
Greenwood Guides to Science and Religion**

Science and Religion, 1450–1900: From Copernicus to Darwin
Richard G. Olson

Science and Religion, 400 B.C. to A.D. 1550: From Aristotle to Copernicus
Edward Grant

Science and Nonbelief
Taner Edis

Judaism and Science: A Historical Introduction
Noah J. Efron

Science and Islam

MUZAFFAR IQBAL

Greenwood Guides to Science and Religion
Richard Olson, Series Editor

Greenwood Press
Westport, Connecticut • London

Library of Congress Cataloging-in-Publication Data

Iqbal, Muzaffar.
　Science and islam / Muzaffar Iqbal.
　　p. cm. — (Greenwood guides to science and religion)
　Includes bibliographical references and index.
　ISBN-13: 978–0–313–33576–1 (alk. paper)
　ISBN-10: 0–313–33576–1 (alk. paper)
1. Islam and science.　I. Title.
BP190.5.S3I672　2007
297.2'65—dc22　　　2007000423

British Library Cataloguing in Publication Data is available.

Copyright © 2007 by Muzaffar Iqbal

All rights reserved. No portion of this book may be
reproduced, by any process or technique, without the
express written consent of the publisher.

Library of Congress Catalog Card Number: 2007000423
ISBN-13: 978–0–313–33576–1
ISBN-10: 0–313–33576–1

First published in 2007

Greenwood Press, 88 Post Road West, Westport, CT 06881
An imprint of Greenwood Publishing Group, Inc.
www.greenwood.com

Printed in the United States of America

The paper used in this book complies with the
Permanent Paper Standard issued by the National
Information Standards Organization (Z39.48–1984).

10　9　8　7　6　5　4　3　2　1

The author and the publisher gratefully acknowledge permission to excerpt material from
the following sources:

Ya'qub ibn Ishaq al-Kindi, *On First Philosophy (fi al-Falsafah al-Ulla)*. Translated by Alfred
L. Ivry. Albany: State University of New York Press, 1974, pp. 70–75. Reprinted by
permission of the State University of New York Press. All rights reserved.

Ibn Sina-al-Biruni Correspondence, *al-As'ilah wa'l-Ajwibah*. Translated by Rafik Berjak and
Muzaffar Iqbal, *Islam and Science*, Vol. 1, Nos. 1 and 2, 2003, pp. 91–98 and 253-260. Courtesy
of the Center for Islam and Science.

Abu Ali al-Husayn ibn Abd Allah Ibn Sina, *The Canon of Medicine (al-Qanun fi'l tibb)*.
Adapted by Laleh Bakhtiar from the translation by O. Cameron Gruner and Mazar H. Shah.
Chicago: Kazi Publications, 1999, pp. 11–14. Courtesy of the Great Books of the Islamic
World, Kazi Publications.

Abu Bakr Muhammad bin Tufayl, *Hayy ibn Yaqzan*, ed. Léon Gauthier, Beirut, 1936,
pp. 70–78, especially translated for the Center for Islam and Science by Yashab Tur.

Every reasonable effort has been made to trace the owners of copyrighted materials in this
book, but in some instances this has proven impossible. The author and publisher will be
glad to receive information leading to more complete acknowledgments in subsequent
printings of the book and in the meantime extend their apologies for any omissions.

Contents

Series Foreword		vii
Preface		xv
Acknowledgments		xxiii
Chronology of Events		xxv
Chapter 1	Introduction	1
Chapter 2	Aspects of Islamic Scientific Tradition (the eighth to the sixteenth centuries)	9
Chapter 3	Facets of the Islam and Science Relationship (the eighth to the sixteenth centuries)	29
Chapter 4	The Mosque, the Laboratory, and the Market (the eighth to the sixteenth centuries)	61
Chapter 5	Islam, Transmission, and the Decline of Islamic Science	103
Chapter 6	Islam and Modern Science: The Colonial Era (1800–1950)	131
Chapter 7	Islam and Modern Science: Contemporary Issues	159
Primary Sources		189
Annotated Bibliography		213
Index		227

Series Foreword

For nearly 2,500 years, some conservative members of societies have expressed concern about the activities of those who sought to find a naturalistic explanation for natural phenomena. In 429 BCE, for example, the comic playwright, Aristophanes parodied Socrates as someone who studied the phenomena of the atmosphere, turning the awe-inspiring thunder which had seemed to express the wrath of Zeus into nothing but the farting of the clouds. Such actions, Aristophanes argued, were blasphemous and would undermine all tradition, law, and custom. Among early Christian spokespersons there were some, such as Tertullian, who also criticized those who sought to understand the natural world on the grounds that they "persist in applying their studies to a vain purpose, since they indulge their curiosity on natural objects, which they ought rather [direct] to their Creator and Governor."[1]

In the twentieth century, though a general distrust of science persisted among some conservative groups, the most intense opposition was reserved for the theory of evolution by natural selection. Typical of extreme antievolution comments is the following opinion offered by Judge Braswell Dean of the Georgia Court of Appeals: "This monkey mythology of Darwin is the cause of permissiveness, promiscuity, pills, prophylactics, perversions, pregnancies, abortions, pornography, pollution, poisoning, and proliferation of crimes of all types."[2]

It can hardly be surprising that those committed to the study of natural phenomena responded to their denigrators in kind, accusing them of willful ignorance and of repressive behavior. Thus, when Galileo Galilei was warned against holding and teaching the Copernican system of astronomy as true, he wielded his brilliantly ironic pen and threw down a gauntlet to religious authorities in an introductory letter "To the Discerning Reader" at the beginning of his great *Dialogue Concerning the Two Chief World Systems*:

> Several years Ago there was published in Rome a salutory edict which, in order to obviate the dangerous tendencies of our age, imposed a seasonable silence upon the Pythagorean [and Copernican] opinion that the earth moves. There were those who impudently asserted that this decree had its origin, not in judicious inquiry, but in passion none too well informed. Complaints were to be heard that advisors who were totally unskilled at astronomical observations ought not to clip the wings of reflective intellects by means of rash prohibitions.
> Upon hearing such carping insolence, my zeal could not be contained.[3]

No contemporary discerning reader could have missed Galileo's anger and disdain for those he considered enemies of free scientific inquiry.

Even more bitter than Galileo was Thomas Henry Huxley, often known as "Darwin's bulldog." In 1860, after a famous confrontation with the Anglican Bishop Samuel Wilberforce, Huxley bemoaned the persecution suffered by many natural philosophers, but then he reflected that the scientists were exacting their revenge:

> Extinguished theologians lie about the cradle of every science as the strangled snakes beside that of Hercules; and history records that whenever science and orthodoxy have been fairly opposed, the latter has been forced to retire from the lists, bleeding and crushed, if not annihilated; scotched if not slain.[4]

The impression left, considering these colorful complaints from both sides, is that science and religion must continually be at war with one another. That view of the relation between science and religion was reinforced by Andrew Dickson White's *A History of the Warfare of Science with Theology in Christendom*, which has seldom been out of print since it was published as a two-volume work in 1896. White's views have shaped the lay understanding of science and religion interactions for more than a century, but recent and more careful scholarship has shown that confrontational stances do not represent the views of the overwhelming majority of either scientific investigators or religious figures throughout history.

One response among those who have wished to deny that conflict constitutes the most frequent relationship between science and religion is to claim that they cannot be in conflict because they address completely different human needs and therefore have nothing to do with one another. This was the position of Immanuel Kant, who insisted that the world of natural phenomena, with its dependence on deterministic causality, is fundamentally disjointed from the noumenal world of human choice and morality, which constitutes the domain of religion. Much more recently, it was the position taken by Stephen Jay Gould in *Rocks of Ages: Science and Religion in the Fullness of Life* (1999). Gould writes:

> I ... do not understand why the two enterprises should experience any conflict. Science tries to document the factual character of the natural world and to develop theories that coordinate and explain these facts. Religion, on the other hand, operates in the equally important, but utterly different realm of human purposes, meanings, and values.[5]

In order to capture the disjunction between science and religion, Gould enunciates a principle of "Non-overlapping magisterial," which he identifies as "a principle of respectful noninterference."[6]

In spite of the intense desire of those who wish to isolate science and religion from one another in order to protect the autonomy of one, the other, or both, there are many reasons to believe that theirs is ultimately an impossible task. One of the central questions addressed by many religions is, what is the relationship between members of the human community and the natural world. This question is a central question addressed in "Genesis," for example. Any attempt to relate human and natural existence depends heavily on the understanding of nature that exists within a culture. So where nature is studied through scientific methods, scientific knowledge is unavoidably incorporated into religious thought. The need to understand "Genesis" in terms of the dominant understandings of nature thus gave rise to a tradition of scientifically informed commentaries on the six days of creation which constituted a major genre of Christian literature from the early days of Christianity through the Renaissance.

It is also widely understood that in relatively simple cultures—even those of early urban centers—there is a low level of cultural specialization, so economic, religious, and knowledge producing specialties are highly integrated. In Bronze Age Mesopotamia, for example, agricultural activities were governed both by knowledge of the physical conditions necessary for successful farming and by religious rituals associated with plowing, planting, irrigating, and harvesting. Thus religious practices and natural knowledge interacted in establishing the character and timing of farming activities.

Even in very complex industrial societies with high levels of specialization and division of labor, the various cultural specialties are never completely isolated from one another and they share many common values and assumptions. Given the linked nature of virtually all institutions in any culture it is the case that when either religious or scientific institutions change substantially, those changes are likely to produce pressures for change in the other. It was probably true, for example, that the attempts of pre-Socratic investigators of nature, with their emphasis on uniformities in the natural world and apparent examples of events systematically directed toward particular ends, made it difficult to sustain beliefs in the old pantheon of human-like and fundamentally capricious Olympian gods. But it is equally true that the attempts to understand nature promoted a new notion of the divine—a notion that was both monotheistic and transcendent, rather than polytheistic and immanent—and a notion that focused on both justice and intellect rather than power and passion. Thus early Greek natural philosophy undoubtedly played a role not simply in challenging, but also in transforming Greek religious sensibilities.

Transforming pressures do not always run from scientific to religious domains, moreover. During the Renaissance, there was a dramatic change among Christian intellectuals from one that focused on the contemplation of God's works to one that focused on the responsibility of the Christian for caring for his fellow humans. The active life of service to humankind, rather than the contemplative life of reflection on God's character and works, now became the Christian ideal for many. As a consequence of this new focus on the active life, Renaissance intellectuals turned away from the then-dominant Aristotelian view of science, which saw the inability of theoretical sciences to change the world as a positive virtue. They replaced this understanding with a new view of natural knowledge, promoted in the writings of men such as Johann Andreae in Germany and Francis Bacon in England, which viewed natural knowledge as significant only because it gave humankind the ability to manipulate the world to improve the quality of life. Natural knowledge would henceforth be prized by many because it conferred power over the natural world. Modern science thus took on a distinctly utilitarian shape at least in part in response to religious changes.

Neither the conflict model nor the claim of disjunction, then, accurately reflect the often intense and frequently supportive interactions between religious institutions, practices, ideas, and attitudes on the one hand, and scientific institutions, practices, ideas, and attitudes on the other. Without denying the existence of tensions, the primary goal of the volumes of this series is to explore the vast domain of mutually supportive and/or transformative interactions between scientific institutions, practices, and knowledge and religious institutions, practices, and beliefs. A second goal is to offer the opportunity to make comparisons across space, time, and cultural configuration. The series will cover the entire globe, most major faith traditions, hunter-gatherer societies in Africa and Oceania as well as advanced industrial societies in the West, and the span of time from classical antiquity to the present. Each volume will focus on a particular cultural tradition, a particular faith community, a particular time period, or a particular scientific domain, so that each reader can enter the fascinating story of science and religion interactions from a familiar perspective.

Furthermore, each volume will include not only a substantial narrative or interpretive core, but also a set of primary documents that will allow the reader to explore relevant evidence, an extensive annotated bibliography to lead the curious to reliable scholarship on the topic, and a chronology of events to help the reader keep track of the sequence of events involved and to relate them to major social and political occurrences.

So far I have used the words *science* and *religion* as if everyone knows and agrees about their meaning and as if they were equally appropriately applied across place and time. Neither of these assumptions is true. Science and religion are modern terms that reflect the way that we in the industrialized West organize our conceptual lives. Even in the modern West, what we mean by science and religion is likely to depend on our political orientation, our scholarly background, and the faith community that we belong to. Thus, for example, Marxists and Socialists tend to focus on the application of natural knowledge as the key element in defining science. According to the British Marxist scholar, Benjamin Farrington, "Science is the system of behavior by which man has acquired mastery of his environment. It has its origins in techniques ... in various activities by which man keeps body and soul together. Its source is experience, its aims, practical, its only test, that it works."[7] Many of those who study natural knowledge in pre-industrial societies are also primarily interested in knowledge as it is used and are relatively open regarding the kind of entities posited by the developers of culturally specific natural knowledge systems or "local sciences." Thus, in his *Zapotec Science: Farming and Food in the Northern Sierra of Oaxaca*, Roberto González insists that

Zapotec farmers ... certainly practice science, as does any society whose members engage in subsistence activities. They hypothesize, they model problems, they experiment, they measure results, and they distribute knowledge among peers and to younger generations. But they typically proceed from markedly different premises—that is, from different conceptual bases—than their counterparts in industrialized societies.[8]

Among the "different premises" is the presumption of Zapotec scientists that unobservable spirit entities play a significant role in natural phenomena.

Those more committed to liberal pluralist society and to what anthropologists like González are inclined to identify as "cosmopolitan science" tend to focus on science as a source of objective or disinterested knowledge, disconnected from its uses. Moreover, they generally reject the positing of unobservable entities, which they characterize as "supernatural." Thus, in an amicus curiae brief filed in connection with the 1986 Supreme Court case that tested Louisiana's law requiring the teaching of creation science along with evolution, for example, seventy-two Nobel laureates, seventeen state academies of science, and seven other scientific organizations argued that

> Science is devoted to formulating and testing naturalistic explanations for natural phenomena. It is a process for systematically collecting and recording data about the physical world, then categorizing and studying the collected data in an effort to infer the principles of nature that best explain the observed phenomena. Science is not equipped to evaluate supernatural explanations for our observations; without passing judgement on the truth or falsity of supernatural explanations, science leaves their consideration to the domain of religious faith.[9]

No reference whatsoever to uses appears in this definition. And its specific unwillingness to admit speculation regarding supernatural entities into science reflects a society in which cultural specialization has proceeded much farther than in the village farming communities of southern Mexico.

In a similar way, secular anthropologists and sociologists are inclined to define the key features of religion in a very different way than members of modern Christian faith communities. Anthropologists and sociologists focus on communal rituals and practices that accompany major collective and individual events—plowing, planting, harvesting, threshing, hunting, preparation for war (or peace), birth, the achievement of manhood or womanhood, marriage (in many cultures), childbirth, and death. Moreover, they tend to see the major consequence of religious practices as the intensification of social cohesion. Many Christians, on the other hand, view the primary goal of their religion as personal salvation, viewing society as at best a supportive structure and at worst a distraction from their own private spiritual quest.

Thus, science and religion are far from uniformly understood. Moreover, they are modern Western constructs or categories whose applicability to the temporal and spatial "other" must always be justified and must always be understood as the imposition of modern ways of structuring institutions, behaviors, and beliefs on a context in which they could not have been categories understood by the actors involved. Nonetheless it does seem to us not simply permissible, but probably necessary to use these categories at the start of any attempt to understand how actors from other times and places interacted with the natural world and with their fellow humans. It may ultimately be possible for historians and anthropologists to understand the practices of persons distant in time and/or space in terms that those persons might use. But that process must begin by likening the actions of others to those that we understand from our own experience, even if the likenesses are inexact and in need of qualification.

The editors of this series have not imposed any particular definition of science or of religion on the authors, expecting that each author will develop either explicit or implicit definitions that are appropriate to their own scholarly approaches and to the topics that they have been assigned to cover.

Richard Olson

NOTES

1. Tertullian. 1896–1903. "Ad nationes," in Peter Holmes, trans., *The Anti-Nicene Fathers*, ed. Alexander Roberts and James Donaldson, vol. 3 (New York: Charles Scribner's Sons), p. 133.

2. Christopher Toumey, *God's Own Scientists: Creationists in a Secular World*. (New Brunswick, NJ: Rutgers University Press, 1994), p. 94.

3. Galileo Galilei, *Dialogue Concerning the Two Chief World Systems: Ptolemaic and Copernican* (Berkeley and Los Angeles: University of California Press, 1953), p. 5.

4. James R. Moore, *The Post-Darwinian Controversies: A Study of the Protestant Struggle to Come to Terms with Darwin in Great Britain and America, 1870–1900*. (Cambridge: Cambridge University Press, 1979), p. 60.

5. Stephen Jay Gould, *Rocks of Ages: Science and Religion in the Fullness of Life* (New York: The Ballantine Publishing Group, 1999), p. 4.

6. Ibid., p. 5.

7. Benjamin Farrington, *Greek Science* (Baltimore: Penguin Books, 1953).

8. Roberto Gonzales, *Zapotec Science: Farming and Food in the Northern Sierra of Oaxaca* (Austin: University of Texas Press, 2001), p. 3.

9. 72 Nobel Laureates, 17 State Academies of Science and 7 Other Scientific Organizations. Amicus Curiae Brief in support of Appelles Don Aguilard et al. v. Edwin Edwards in his official capacity as Governor of Louisiana et al. (1986), p. 24.

Preface

The story of the relationship between Islam and science can be told from a variety of perspectives, ranging from the sociological to the historical and from the metaphysical to the scientific. The methodology used for this narrative depends, to a large extent, on how one perceives the *nature* of the relationship between Islam and science. It is, therefore, important to state at the outset the way in which this story will be told in this book.

This question of perspective and methodology has become even more important in recent years, because the enormous amount of theoretical work published by scholars working in the field of science and Christianity in the West has established a certain model for exploring issues related to the interaction between science and religion, and this model seems to have gained general acceptability. In this model, which can be called the "two-entity model," science and religion are taken as two separate entities. These two definitively distinct entities are then posited against each other and are allowed a variety of possible modes of interaction, such as "conflict," "independence," "dialogue," and "integration" (Barbour 2000). This variety, however, is *within* the two-entity model; in other words, these ways of explaining the relationship between science and religion all assume that "science" and "religion" are two separate entities that have a finite number of possible modes of interaction. Each of these modes can be further subdivided into various possibilities, refined, classified, and graded in terms of the degree of interaction being strong or weak, but the model itself remains anchored in the foundational paradigm that considers the two phenomena separate and distinct entities.

The two-entity model has evolved from a specific cultural, historical, and scientific background, and it is supported by episodes from the history of interaction between science and Christianity in the Western world. It is, however, now being claimed that this model is universal, and can be used to understand the relationship between all scientific traditions and all religions (Barbour 2002). While this model has been criticized for certain shortcomings, this criticism has for the most part itself remained within the two-entity framework (Cantor and Kenny 2001). Since science as we understand it today is generally taken to be that enterprise borne of the European Scientific Revolution of the seventeenth century, and since this particular scientific tradition has had a series of conflicts with Christianity, the "conflict model" has gained credibility both in the scholarly world as well as in the popular mind. Furthermore, since the science begotten by the European Scientific Revolution has now spread to all corners of the world, the particular history of the interaction of science and the faith also seems to have accompanied this spread of science: the only adjustment deemed necessary is the substitution of Christianity with Islam or other religions of the world.

This model is being applied retroactively to the historical interaction between various religions and philosophies—for example, the Greek, Roman, and Islamic scientific traditions. It has thus become popular to search out, for example, instances of conflict or cooperation between Islam and the scientific tradition that emerged in Islamic civilization. This approach makes no distinction between premodern and modern science as far as their philosophical foundations are concerned, despite the basic differences between the worldviews that gave birth to the two scientific traditions.

Given this background, we must first ask a basic question: is the two-entity model—arising out of a particular cultural, historical, and scientific background—truly applicable to all religions and all scientific traditions?

This model can only be applicable to all religious and scientific traditions if

1. nature—the subject matter of science—and its relationship with both God and humanity is understood in the same way in all religious traditions;
2. the foundational source texts of all religions are parallel to the Bible in the epistemological, metaphysical, and semantic structures they imply; and
3. science in all civilizations is an enterprise that has remained the same over centuries as to its foundations.

What we have, however, is a situation where even the term *science* has no acceptable universal definition (Ratzsch, 2000, 11). Furthermore, neither nature nor science nor their mutual relationships are conceived uniformly across religious traditions or even within a single tradition over centuries. The two-entity model becomes especially problematic when we take into account developments in the history of philosophy of science. Most philosophers of science agree that what we now term science is not a label that can be evenly applied to the investigation of nature in all eras and all civilizations: that is, the term *science* (however one defines it) rests upon a number of conceptual presuppositions peculiar to that civilization, formed by the social, cultural, and historical ethos of that specific civilization. What is known as science today is generally understood to be that enterprise begun in the seventeenth century, built on the spectacular experiments and theories of Galileo (1564–1642), Kepler (1571–1630), and Newton (1642–1727), and later entrenched in the social, economic, academic, and cultural institutions of Western civilization. In fact, the emergence of this particular scientific tradition has changed the very concept of science as it existed prior to the seventeenth century. Many contemporary historians of science tend to term scientific activity prior to the emergence of the seventeenth century and dating back two thousand years to Greece as "Natural Philosophy" rather than science.

Furthermore, the particular history of the relationship between modern science and Christianity is especially inapplicable to Islam, since there is no church in Islam, no ecclesiastic authority that could have entered into interaction with scientists and scientific institutions in any formalized concrete form. There is no Pope who could have established the "Holy Office of the Inquisition," as Pope Paul III did in 1542, or who could have issued an Index of Prohibited Books or encyclics on Islam and science. This is not to say that Islamic religious thought is without its own internal conflicts—and there will be much mention of these in subsequent chapters—but to underscore the reasons why the two-entity model, often used to describe the interaction of Christianity and science, is inapplicable to the case of Islam (and possibly other religious traditions).

There is no known scientist or religious scholar between the eighth and seventeenth centuries who felt the need to describe the relationship between science and religion by composing a book on the subject. This absence of "Islam and science" as a differentiated discipline is proof in itself that during these long centuries—when the Islamic scientific tradition was the world's most advanced enterprise of science—no one felt the need to relate the two through some external construct. When the need did arise, it arose in an entirely different context and in an entirely different era—that is, with the arrival of modern science in the lands of Islam during colonization of the Muslim world.

As mentioned, Islam does not view nature as a self-subsisting entity that can be studied in isolation from its all-embracing view of God, humanity, and the cosmological setting in which human history is unfolding. Furthermore, in Islamic classification of knowledge, science—the discipline that studies nature—is taken as but one branch of knowledge, integrally connected with all other branches of knowledge, all of which are linked to the concept of *Tawhid*, the Oneness of God. Thus, the Islamic tradition does not regard any discipline of knowledge independent of other disciplines. For this reason, the connector "and" in the phrase "Islam and science" does not attempt to connect two separate entities, rather, it is used here as a copula. There is also a fundamental difference between the nature of science that existed in Islamic civilization between the eighth and sixteenth centuries and modern science; they approach nature in two very different manners and hence one cannot use the same methodology for narrating the stories of the interaction of Islam with both premodern and modern science.

This division is not arbitrary but is borne out of the subject matter itself: the science prior to the arrival of modern science in the Muslim world had its own distinct mode of interaction "with" Islam because it arose from within the greater matrix of Islamic civilization. Of course, these ways of interaction did not exist in vacuum; they existed within the larger intellectual universe of Islam—a universe that had its share of sharp edges and discordant voices, but which was, nonetheless, a universe shaped by the Qur'anic worldview.

With the arrival of modern Western science in the Muslim world, Islam and science discourse entered a new period. Because this arrival coincided with the colonization of the Muslim world it was accompanied by numerous other factors, including the economic, political, and military agendas of the colonizing powers. This destroyed the institutions that had produced the eight-hundred-year-old Islamic educational tradition. The strangulation of centuries-old institutions and the implantation of new scientific institutions with agendas that suited the interests of the colonizing powers changed the dynamics of the practice of science in the Muslim world. Now Islam had to interact with a science based on a philosophy of nature foreign to its own conception. In order to explain this new relationship we will have to take into consideration certain foundational epistemological assumptions of modern science, as well as developments that led to the emergence of these new concepts of nature in the post-Baconian Western world, and see how these radical changes shaped the discourse on Islam and science. Also important for our purpose is the complex process of intellectual colonization of the Muslim mind, which produced a deep-seated inferiority complex with respect to Western science and technology—the factors perceived as the main reasons for the West's domination of the world and the colonization of Muslim lands.

Finally this book will discuss the post–World War II era, which has produced a certain degree of clarity in the discipline and which promises to open new vistas in the future. The current fervor of intellectual activity in the Muslim world—as it reshapes and reconfigures in a world largely constructed by Western science and technology—is accompanied by a tremendous amount of intellectual and physical violence and chaos, but such ataxic disorder is not new to the Islamic tradition. There have been several such periods in history when Muslims were forced to reshape their intellectual tradition, physical borders, and their relationships with other communities and traditions. Some of these periods were particularly stark and accompanied by much bloodshed, others produced intellectual displacements, but, throughout the last fourteen hundred years, the Islamic tradition has remained resilient and able to cope with apparently insurmountable obstacles. Whether it was the intellectual encounter with the legacy of Greek philosophy and science—an encounter we will discuss in detail in the next chapter—or the large-scale physical destruction caused by marauding tribes sweeping from the steppes of Central Asia, Islamic tradition was always able to reemerge with renewed vigor.

The current Islam and science discourse needs to be viewed in the larger context of the encounter of Islam with modernity and its intellectual, social, cultural, political, and economic outlooks, and so the last part of our story has to be narrated within this broader setting. Of course, it remains to be seen how Islamic tradition will fare in this new encounter, as this is a situation of another kind; modern science and technology are rapidly reshaping the entire spectrum of human existence—from the way human beings are born to the way they procure their food, travel, communicate, establish interpersonal relationships, and die. To be sure, it is a fascinating story that deserves our full attention, as the world around us reshapes through encounters of a kind never before witnessed in human history.

The questions we wish to explore in the following chapters include the following: What was Islamic in Islamic science? How did Islam affect the course of development of the Islamic scientific tradition? Were there any tensions within the Islamic tradition that may have inhibited the full blossoming of this scientific activity? What were the distinct contributions of the Islamic scientific tradition to broader scientific knowledge? When, why, and how did this tradition come to an end? How was this scientific knowledge passed on to Europe? What are the new facets of the Islam and science relationship, which have appeared in the post–Scientific Revolution era? What are some of the fundamental issues in the contemporary Islam and science discourse?

Through an exploration of these and related questions this book attempts to present a spectrum of Islamic opinions on some of the most important questions in the religion and science discourse.

A NOTE ON NAMES, DATES, TRANSLATED TEXTS, AND TECHNICAL TERMS

The Arabic word Allah is used whenever it is necessary to mention God by His personal name. It is customary to capitalize the recognized "fair names of Allah" and pronouns denoting God. It is also customary to capitalize the word "companion(s)" when used for those men and women who were the Companions of the Prophet of Islam, whose own mention is customarily followed by salutation by Muslims. This book follows all these, but the last-mentioned, as it is generally absent from works such as this with the proviso that readers should nevertheless send the customary salutations on the Prophet whenever his name is mentioned. Passages of the Qur'an cited in the text are italicized; all translations are mine, though necessarily based on a number of other translations and commentaries. All references to the Qur'an in the text occur within parentheses () and are prefixed with "Q." and two numbers separated by a colon, of which the first number refers to the chapter and the second to the verse.

Transliteration has been kept to a minimum. A distinction is, however, made between the Arabic letter *hamzah*, represented by an apostrophe ('), and the letter *'ayn*, represented by a single beginning quotation mark ('). Where absolutely necessary, the letters "â," "î," and "û" have been used to indicate long vowels, and a circumflex ("^") on top of the letter to distinguish two similar sounding letters (e.g., "š" to distinguish "*šâd*" from "*sîn*").

The names of many Muslim scientists and scholars of premodern times have been corrupted for various reasons, a tradition stemming from medieval translations. Al-Ghazali, for instance, became Algazel; Abu Ma'shar became Albumasar; Ibn Tufayl became Abubacer; al-Bitruji became Alpetragius; al-Farghani became Alfraganus; Razi became Rhazes; Abu'l-Qasim al-Zahrani became Albucasis; and Ibn Sina became Avicenna. Unfortunately, this practice, which first emerged at a time when Arabic names were either carelessly Latinized or were simply distorted because of a lack of proper linguistic skills, continues in many works that do not use accurate transliteration schemes. For reasons of expediency, this book also does not use a transliteration system for Arabic words, but it does preserve original names.

Arabic names can be confusing for readers unfamiliar with this system. Many names cited in this book are not really the personal names (such as Muhammad, Husayn, or Ali) of the scientist or scholar because in most cases they are known through family associations (as in "ibn Sina," the son of Sina) or simply by alluding to their tribe, place, or region of birth (as in "al-Khwarizmi," meaning "of or from Khwarzam"). Sometimes a person is known by his paternal name, *kunya* (Abu Hamid, "the father of Hamid").

A nickname, an occupational name, or a title can also become the most well known name of a person (al-Jahiz, "the goggle-eyed"; al-Khayyam, "the tent maker"). In this book, the full name is given at first use and the most well known name is used thereafter.

Another point of import is the pronoun used by the writer. In Islamic tradition, the plural first-person "we" rather than "I" is preferred, for a variety of reasons. "I" is considered impolite because of the emphasis this pronoun has on the individual, whereas "we" takes the emphasis away from the single person—the additional persona is left ambiguous, but it is assumed that the writer is not alone in the process of writing; that he or she is being aided by others; that the Divine presence is there in the very act of transmission of knowledge and ideas; that a whole host of scholars are in company. For these and other reasons we seldom find the use of "I" in Islamic scholarly tradition, and this book follows precedent.

For reasons of simplicity, only one system of dates is used. Unless specified, all dates refer to Common Era. Dates of birth or death are mentioned when a person's name first appears in the text; "*ca.*" is used to indicate approximate date.

Acknowledgments

Like all books, *Science and Islam* has come into existence through the efforts of numerous individuals who participated in its publication in various capacities. This participation ranges from the most practical task of copy editors who improved its manuscript to men and women of bygone eras who contributed to the making of the Islamic scientific tradition. Then there are the historians of that tradition who, over the last three centuries, have unearthed countless manuscripts, instruments, and books which allow us to reconstruct the contours of the enterprise of science in Islamic civilization prior to the emergence of modern science; without their painstaking efforts we would not know what we do now about this tradition which was until recently considered a mere appendix to the Greek science.

Recent trends in telling the story of Islamic scientific tradition are revisionist in nature, attempting a major reconstruction of the narrative that dominated the field until the close of the twentieth century. This book attempts to highlight these efforts aimed at a more accurate description of science in Islamic civilization. As such, I am particularly indebted to those historians who are at the forefront of this effort.

Any new work on the relationship between Islam and science is bound to raise more questions than it can answer. The complex matrix from which this relationship emerges involves history, politics, economics, sociology, and, of course, subhistories of various branches of science. Any conclusions drawn in a work of this nature are bound to remain provisional until more information becomes available about the scientific, social, political, and cultural conditions of those centuries during which Islamic scientific enterprise flourished in lands as far apart as Central Asia and al-Andalus.

But even when this information becomes more accessible, there will always be disagreement among scholars about how to interpret data, and this book highlights a number of such disagreements. I, nevertheless, appreciate the work of those scholars with whom I have disagreed in certain fundamental ways, as they have contributed to the discourse just as those whose opinions, interpretations, and findings are in consonance with my own views.

In writing this book I have benefited from the work of scholars in such diverse fields as the history of science, Qur'anic exegesis, *hadith* studies, general history of Islam, sociology, linguistics, and philosophy of science, to name only a few. Their rich contributions have greatly enhanced our understanding of the various aspects of Islamic civilization, including the enterprise of science.

In particular, I am thankful to Richard Olson, series editor, for inviting me to contribute this volume to the Greenwood Guides to Science and Religion, and to Kevin J. Downing of Greenwood Press, whose promptness in responding to numerous queries during the course of the writing of this work has been greatly helpful. Thanks are also due to the copyright owners of primary source material and images used in this book; they are more fully acknowledged in the text itself.

Most of all, I am thankful to members of my immediate family: Basit Kareem Iqbal, my dear son, whose editing skills have greatly improved this book; my wife Elma, whose sustained involvement in this project has enriched it in numerous ways; to my daughter Noor and son Usman, whose indirect contributions to this project range from the sheer pleasure which their presence evokes to the nurturing of ideas through various discussions. It is because of these contributions that it was possible to devote so much time and effort to this book in the enriching and blissful environment of Wuddistan—a place where it is possible to preclude distractions of the world and devote long hours to a solitary labor of love in the reconstruction of a past which may slip from human memory if not preserved. It is needless to say that only I am responsible for the flaws and shortcomings in this book.

Wuddistan
Dhu'l Qa'dah 12, 1427 AH
December 3, 2006

Chronology of Events

Sixth Century

570–632   Prophet Muhammad, born in Makkah, died in Madinah

Seventh Century

610   Muslims migrated from Makkah to Madinah (This marked the beginning of the Hijri calendar)
632–655   Muslim conquests of Syria, Iraq, and Egypt
661   Establishment of Umayyad dynasty
691   Dome of the Rock built in Jerusalem

Eighth Century

706   Great Mosque of the Umayyads built in Damascus
710   Muslim forces entered Spain
712   Muslims reached Samarqand
       al-Asmai (739–831), the first Muslim scientist known to contribute to zoology, botany, and animal husbandry
750   Establishment of Abbasid dynasty; end of Umayyad dynasty
754–775   The reign of caliph al-Mansur, patron of early translations of scientific texts from Greek
762   Baghdad founded by Abbasids

	Amr ibn Bakr al-Jahiz (779–868), zoologist, lexicographer
785	Mosque of Cordoba begun
796	First university in al-Andalus (Spain)
	Jabir ibn Hayyan (*ca.* late eighth and early ninth centuries), "father of alchemy" who emphasized scientific experimentation and left an extensive corpus of alchemical and scientific works

Ninth Century

ca. 800	Teaching hospital established in Baghdad
	Banu Musa brothers (*ca.* 800–873) supervised the translation of Greek scientific works into Arabic, founded Arabic school of mathematics, wrote over twenty books on subjects such as astronomy, ingenious devices, on the measurement of plane and spherical figures
	Ibn Ishaq al-Kindi (801–873), philosopher, mathematician, specialist in physics, optics, medicine, and metallurgy
	Hunayn ibn Ishaq (809–873) collected and translated Greek scientific and medical knowledge into Arabic
	Abbas ibn Firnas (810–887), technologist and chemist, the first man in history to make a scientific attempt at flying
813–848	Flourishing of Mu'tazilite philosophical theology
832	Establishment of Bayt al-Hikmah library in Baghdad by caliph al-Ma'mun
	Thabit ibn Qurra (835–901), mathematician who worked on number theory, astronomy, and statics
	al-Khwarizmi (b. before 800, d. after 847), mathematician, geographer, and astronomer, the founder of algebra
ca. 850	Arabic treatises written on the astrolabe
	al-Battani (850–922), astronomer who determined the solar year as being 365 days, 5 hours, 46 minutes, and 24 seconds
	Abu Kamil (859–930), mathematician who applied algebraic methods to geometric problems

Chronology of Events xxvii

al-Farghani (d. after 861), astronomer, civil engineer who supervised construction of the Great Nilometer canal of Cairo

Sabur ibn Sahl (d. 868), pharmacist

al-Tabari (b. *ca.* 808 and d. *ca.* 861), physician, natural scientist who wrote on medicine, embryology, surgery, toxicology, psychotherapy, and cosmogony

Abbas ibn Firnas (d. 888) studied mechanics of flight, developed artificial crystals, built a planetarium

Tenth Century

874–999   Samanid dynasty in Transoxania

Ishaq ibn Hunayn (d. 910), physician and scientific translator from Greek and Syriac to Arabic

Abdul Ibn Khuradadhbih (d. 912), geographer who developed a full map of trade routes of the Muslim world

Abu Bakr Zakaria al-Razi (b. *ca.* 854 – d. 925 or 935), philosopher and physician with specialty in ophthalmology, smallpox, and chemistry

al-Battani (b. before 858–d. 929), mathematician of trigonometry and astronomer

al-Farabi (b. *ca.* 870–d. 950), metaphysician, musicologist, philosopher of sociology, logic, political science, and music

al-Sufi (903–986), astronomer renowned for his observations and descriptions of fixed stars

Ibrahim ibn Sinan (908–946), mathematician and astronomer, who studied geometry, particularly tangents to circles, also advanced theory of integration, apparent motions of the sun, optical study of shadows, solar hours, and astrolabes

931   Baghdad initiated a licensing examination for physicians, attended by 869 doctors

al-Zahrawi (936–1013), "father of surgery," wrote a thirty-volume work of medical practice

al-Buzjani (940–998), astronomer and mathematician who specialized in trigonometry and geometry

936   Madinat al-Zahra palace complex built in Cordoba

al-Ash'ari (d. 936), founder of school of Islamic theology

Ibn Yunus (950–1009), astronomer known for his many trigonometrical and astronomical tables

Abu Hasan Ali al-Masudi (d. 956 or 957), geographer and historian who conceived of geography as an essential prerequisite of history

al-Khazin (d. 971), astronomer and mathematician, whose *Zij al-safa'ih* (*Tables of the Disks of the Astrolabe*) were the best in his field

al-Khujandi (d. 1000), mathematician and astronomer who discovered the sine theorem relative to spherical triangles

Ibn al-Haytham (965–1040), astronomer, mathematician, optician, and physicist

973   Al-Azhar University founded in Cairo

al-Baghdadi (980–1037), mathematician, wrote about the different systems of arithmetic

al-Jayyani (989–1079), mathematician, wrote commentaries on Euclid's *Elements* and the first treatise on spherical trigonometry

969–1171   Fatimid dynasty in Egypt

Eleventh Century

al-Baqilani (d. 1013), theologian who introduced the concepts of atoms and vacuum into the Kalam, extending atomism to time and motion, conceiving them as essentially discontinuous

Ibn al-Thahabi (d. 1033), physician, who wrote the first known alphabetical encyclopedia of medicine with lists of the names of diseases, medicines, the physiological process or treatment, and the function of the human organs

Abu Rayhan al-Biruni (973–1048), one of the greatest scientists in history, astronomer and mathematician who determined the earth's circumference

Ibn Sina (980–1037), the greatest physician and philosopher of his time, the author of *Canon of Medicine* al-Zarqali (1028–1087), mathematician and astronomer, who excelled at the construction of precision instruments for astronomical use; constructed a flat astrolabe that was "universal" for it could be used at any latitude; built a water clock capable of determining the hours of the day and night and indicating the days of the lunar months

1085	Christians took Toledo

Ibn Zuhr (1091–1161), physician, surgeon, most prominent parasitologist of the Middle Ages, the first to test different medicines on animals before using them with humans

1095	Pope Urban called for Crusade
1099	Jerusalem taken by Crusaders

Twelfth Century

Omar Khayyam (1048–1131), mathematician, astronomer, and philosopher, well known in the West for his poetry

Abu Hamid al-Ghazali (1058–1111), philosopher and theologian, the most influential scholar, who remains an important intellectual figure to this day

Ibn al Tilmidh (b. *ca.* 1073–1165), pharmacist who wrote *al Aqrabadhin*

Ibn Bajjah (d. *ca.* 1138), philosopher, scientist, physician, musician, commentator on Aristotle

al-Idrisi (1100–1166), geographer and cartographer, who constructed a world globe map of 400 kg pure silver, precisely recording on it the seven continents with trade routes; his work was the best of Arab-Norman scientific collaboration

Ibn Tufayl (1105–1186), Andalusian physician, author of *Hayy bin Yaqzan*

Gerard of Cremona (b. *ca.* 1114–1187), translator of works on science from Arabic to Latin

al-Khazini (b. *ca.* 1115–d. *ca.* 1130), astronomer, engineer, inventor of scientific instruments, wrote about the science of weights and art of constructing balances, renowned for his hydrostatic balance

Ibn Rushd (1126–1198), Andalusian physician, philosopher, judge, astronomer, commentator on Aristotle; the most popular Muslim philosopher in the Latin West

al-Samawal (1130–1180), mathematician, extended arithmetic operations to handle polynomials

al-Bitruji (*fl. ca.* 1190), astronomer, natural philosopher, and mathematician

Crusaders defeated by Salah al-Din Ayyubi (d. 1193)

al-Suhrawardi (d. 1191), one of the greatest mystics of Islam who integrated philosophical cosmologies with Islamic worldview

Thirteenth Century

Ibn al-Baytar (b. *ca.* 1190–d. 1248), pharmacologist and botanist who systematized the known animal, vegetable, and mineral medicines

al-Jazari (*fl. ca.* 1204), inventor of technologies, wrote *The Book of Knowledge of Ingenious Devices*

Fahkr al-Din al-Razi (d. 1210), exegete and philosopher, author of the commentary *Mafatih al-Ghayb*

al-Samarqandi (d. 1222), physician who wrote on symptoms and treatment of diseases

Ibn Nafis (1213–1288), physician who first described the pulmonary circulation of the blood, whose *Comprehensive Book on the Art of Medicine* consisted of 300 volumes

Ibn Jubayr (d. 1217), philosopher, geographer, and traveler

1219–1329   Mongol raids caused enormous destruction of cities and economies

al-Maghribi (1220–1283), trigonometrist and astronomer

1236   Christians took Cordoba

Ibn al-Arabi (d. 1240), exponent of Sufi monism, wrote important works on cosmology

al-Suri (d. 1241), pharmacologist who documented every known medicinal plant

Chronology of Events xxxi

1258	Baghdad destroyed by Mongols, ended the Abbasid Caliphate
1261	Mamluk dynasty established in Egypt
	Nasir al-Din al-Tusi (1201–1274), astronomer and mathematician who specialized in non-Euclidean geometry
	al-Qazwini (b. *ca.* 1203–1283), cosmographer and geographer
1232–1492	Construction of al-Hambra
1236	Fall of Cordoba to Christians
1244	Crusaders lost Jerusalem for the last time
1254–1517	Mamluk rule in Egypt
1261	Foundation of the *gazi* principalities in Western Anatolia, the beginning of the Ottoman Empire
1258	Baghdad taken by Mongols
1267	First Muslim state of Samudra Pasai in Indonesia
1271	Latin empire collapsed; Byzantine rule restored in Constantinople
	Jalal ud-Din Rumi (d. 1273), Sufi, the author of *Masnavi*

Fourteenth Century

1300	Timbuctu, Mali, and Gao became important centers of learning
1300–1453	Growth of Ottoman Empire
1303	Mongols defeated by Mamluks in Egypt
	Ibn al-Shatir (1304–1375), Damascene astronomer who developed the concept of planetary motion and astronomically defined the times of prayer
	Ibn al-Banna (d. 1321), poet, scholar
	al-Jawziyyah (d. 1350), Qur'an scholar
	Ibn Battutah (1304–d. 1368 or 1377), famed traveler, covered 75,000 miles in 22 years across Europe, Africa, the Middle East, and Asia, made extensive contributions to geography
1308–1312	Mali sultans traveled across the Atlantic, explored the lands around the Gulf of Mexico and the American interior via the Mississippi River

al-Khalili (1320–1380), astronomer who compiled extensive tables for astronomical use

Kamaluddin Farsi (d. 1320), astronomer who improved Ibn al-Haytham's *Optics*, studied reflection and the rainbow

Ibn Khaldun (1332–1406), philosopher of history and society, wrote a universal history *Muqaddimah*

Qutb al-Din Shirazi (1236–1311), astronomer who perfected Ptolemaic planetary theory

ca. 1313   Golden Horde Mongol khans converted to Islam

Izz al Dinn al Jaldaki (d. 1360), chemist

Ulugh Beg (1394–1449), ruler of Samarqand and an accomplished astronomer; his astronomical observatory completed in 1429

Fifteenth Century

Timur (d. 1405) ruled from the Ganges to the Mediterranean

al-Umawi (1400–1489), mathematician who wrote works on arithmetic and mensuration

al-Qalasida (1412–1486), mathematician who introduced ideas of algebraic symbolism by using letters in place of numbers

Ghiyath ud-Din Kashi (d. 1429), astronomer and mathematician who worked at Samarqand Observatory

1453   Ottomans took Constantinople

1455   Papal Bull authorized Roman Catholics to "reduce to servitude all infidels"

1492   Christians took Granada; Muslim states ended in Spain; more than a million volumes of Muslim works on science, philosophy, and culture were burnt in the public square of Vivarrambla in Granada

1498   Vasco Da Gama in India with the help of the Arab Muslim navigator Ahmad Ibn Majid (b. 1432) who wrote the encyclopedic *Kitab al-Fawa'id fi Usul 'Ilm al-Bahr wa'l-Qaw'id* (*Book of Useful Information on the Principles and Rules of Navigation*)

Chronology of Events xxxiii

Sixteenth Century

ca. 1500–1700	The height of power of the three post-Abbasid empires:
	(a) Ottoman Empire (1299–1924)
	(b) Safavid Empire (1501–1736)
	(c) Mughal Empire (1526–1857)
1528	Sankore University founded in Timbuktu

Seventeenth Century

1602	Dutch East India Company founded
	Mulla Sadra (d. 1641), the most important Muslim philosopher of the post-Ibn Sina era, the author of the *Four Intellectual Journeys*
1683	Siege of Vienna by Turks
1688	Fall of Belgrade to Ottomans

Eighteenth Century

1781–1925	Qajar Empire in Iran
	Aurangzeb (d. 1707), Mughal emperor
1722	Afghans conquered Iran and ended the Safavid dynasty
1736	Nadir Shah (d. 1747) assumed the title of Shah of Persia and founded the Afsharid dynasty, defeated the Mughals and sacked Delhi 1739
1757	Battle of Plassey marked the beginning of British rule in India
1789	Napoleon arrived in Egypt
	Hajji Mulla Hadi Sabziwari (b. 1797 or 1798–1873), philosopher, most prominent of the Qajar period

Nineteenth Century

Most of the Muslim world colonized by European powers

Syed Ahmad Khan (1817–1898) published an incomplete scientific commentary (*tafsir*) on the Qur'an

	Sayyid Jamal al-Din al-Afghani (1838–1897), Muslim thinker who debated with the French philosopher Ernst Renan on religion and science
	New scientific institutions modeled on European patterns were established in the colonies; official languages changed to English or French in most of the Muslim world
1857	India became a British colony
	Muhammad ibn Ahmad al-Iskandarani, an Egyptian physician published the first complete scientific *tafsir* of the Qur'an in 1880

Twentieth Century

1900–1920	Nationalistic movements emerged in many parts of the Muslim world
1920–1950	Struggle for independence throughout the Muslim world
1950–1965	Decolonization and rapid emergence of new nation states in the entire Muslim world
1950–1980	Large numbers of Muslim students and emigrants arrived in the West
1975	After the dramatic rise in oil prices, many oil-rich Muslim countries imported Western science and technology; new scientific institutions emerged in the Muslim world
1975–2000	Islam and Science discourse matures. A number of new perspectives emerge
1980–1990	Some Muslim countries established institutions for research on the scientific verses of the Qur'an, held conferences and published books on Islam and science
1980–2000	Teaching of modern science at postgraduate level was given greater attention and state patronage in certain Muslim countries such as Pakistan, Saudi Arabia, Malaysia, Iran, Egypt, and Turkey
1988–2000	Following Gulf War I, many Muslim states made efforts to acquire modern defense technology
1990–2000	Exponential increase in the number of Muslim websites dealing with Islam and science
1998	Pakistan successfully tested a nuclear weapon

Twenty-first Century

Rapid spread of new technologies and consumer goods, such as cell phones, throughout the Muslim world

Following the September 11, 2001, terrorist attacks on the United States, many Western countries imposed restrictions on Muslim students entering certain fields of science and technology.

Increased Muslim presence on the Internet, more websites making claims about the presence of modern scientific theories and facts in the Qur'an; simultaneous maturity of Islam and science discourse as more scholars pay attention to the fundamental issues in the discourse

Chapter 1

Introduction

This book narrates the story of the interaction between Islam and science, from the time of Islam's emergence in the seventh century to the present. This story is divided into two broad periods: the era prior to the emergence of modern science and from the seventeenth century to the present. This division is not arbitrary; rather, it arises from the very nature of the scientific enterprise, which has gone through a foundational change since the Scientific Revolution of the seventeenth century. Before we embark on this exploration of the relationship between Islam and science, however, something must be said about Islam itself, for—despite its ubiquitous presence in the media—it remains one of the most poorly understood religions in the world. Along with contemporary events contributing to the general perception of Islam in the West, there is also silent historical baggage underlying non-Muslim perspectives on the religion, making the understanding of Islam a complicated affair. The span of the twentieth century reveals a phenomenal shift: at its dawn, Islam was considered by many to have lost its appeal, or at least was a religion without a definable polity, as most of its adherents then lived in colonies occupied and ruled by European powers. At the dawn of the twenty-first century, the word *Islam* has become commonplace and hardly a day passes without mention of Islam and Muslims in headline news. This enormous attention has, however, not produced an understanding of the religion based on its own sources; extensive journalistic coverage of events has been responsible instead for distortions of the message of Islam. It is, therefore, necessary to begin with a brief account of the emergence of Islam in the seventh century in Makkah, a remote town in Arabia.

THE EMERGENCE OF ISLAM

Makkah, where Muhammad, the future Prophet of Islam, was born, was not located on the crossroads of any major civilization of the time. It was nevertheless a town of considerable importance owing to the presence of the Ka'bah, a Sanctuary built there by Abraham and his son Ishmael some 2500 years before the birth of Muhammad to Aminah on a Monday in the month of April, 571. Her husband, Abdullah, had died a few months before the birth of the child while returning home from Syria, where he had traveled with one of the trade caravans that left Makkah twice a year—for Yemen in winter and for Syria in summer. Shortly after his birth, the child was taken to the Ka'bah and named Muhammad by his grandfather, Abdul Muttalib, a descendent of Abraham through his son Ishmael.

At the age of forty, when Muhammad had been married to a wealthy widow of Makkah for almost fifteen years, he received his first revelation while in retreat in a cave at the summit of the Mountain of Light, about five kilometers south of the ancient Sanctuary: *Recite! Recite in the name of thy Sustainer Who created; Created Man from a clot of blood; Recite—for thy Sustainer is the Most Bountiful—Who taught by the pen; taught Man what he knew not* (Q. 96:1-5). This was the beginning of the descent of the Qur'an—the Book which is at the heart of all things Islamic.

The revelation continued over the next twenty-three years, culminating a few days before the death of the Prophet in Madinah, an oasis on the caravan route to Syria, some four hundred kilometers north of Makkah and to where the Prophet had migrated in September 622. This migration, called the *Hijrah*, marks the beginning of the Islamic calendar as well as the establishment of the first Islamic state. At the time of the Prophet's death in June 632 at the age of sixty-three, this state had expanded to encompass all of Arabia, and an expedition was on its way to Syria, at the northern border of the nascent state. Within a century of his death, Islam had spread throughout the Middle East, much of Africa and Asia, and as far as Narbonne in southern France. This expansion of the geographical boundaries of the Muslim world occurred in two successive waves. The first took place between 632 and 649 and the second between 693 and 720.

This rapid expansion of the geographical boundaries of the Islamic state has often led Western historians to view Islam as a religion that spread by the sword. This image of Islam was constructed during the Middle Ages against the background of the Crusades and has remained entrenched in the West to this day (Grant 2004, 231). As far as Muslims are concerned, it was the inherent truth of the message of Islam, rather than military victories, that established Islam across this vast region, extending from Makkah to southern France on the one hand and from the steppes of Central Asia to the barren deserts of Africa on the other. A serious inquiry into the rapid spread of Islam during its first century is bound to produce a different answer than brute force, because it defies both logic and historicity that a few Arab armies commanding at most a few thousand soldiers, passing through certain selected routes of the ancient world, could bring such a large number of people into the fold of Islam. In any case, more important for the purpose of our book are two revolutions of another kind: a vast social revolution that took place during this first century of Islam, and an intellectual revolution of the first order that transformed the nomadic Arab society within two generations.

THE TWO REVOLUTIONS

The social revolution was catalyzed by the rapid incorporation of numerous cultures into the fold of Islam. Within those first hundred years, a very large number of micro-transformations took place in the ancient lands of Asia, Africa, and Central Asia to give birth to a new social and cultural mosaic. It is this multicultural, multiethnic, and even multireligious society that was to produce the conditions for the seven-century-long enterprise of science in Islamic civilization. This vast social change in the eighth century was the result of mass conversions, migrations, the establishment of new cities and new administrative and fiscal institutions, and a new social contract that Islam established with Jewish and Christian communities. The belief system of Islam was able to incorporate varied social and cultural traditions because of its overarching universal nature. Thus, whether it was the hunting rites of the nomadic tribes of Central Asia or the rituals of marriage and death of the ancient city dwellers of Mesopotamia, Islam was able to recast these social and cultural aspects of various peoples through reorienting them to its uncompromising monotheism, a process that often left the outer form untouched while removing any trace of polytheism. This allowance encouraged cultural diversity and produced a vast synthesis of numerous cultures over a very large region of the old world.

The expansion of geographical borders and the subsequent cultural synthesis forced the keenest minds of the times to continuously formulate answers to a wide range of questions arising from, among other things, new theological concerns, specific needs of the newly converted masses and immigrants, and the emergence of new administrative and financial arrangements between the state and its citizens. The rapid geographical expansion also posed new questions related to the practice of faith and that required immediate attention: how to determine the correct *qiblah* (direction toward the Ka'bah in Makkah) for the five obligatory daily prayers from a distant city; how to calculate the correct amount of *zakah* (obligatory charity on wealth and material goods) on goods that did not exist in Madinah during the life of the Prophet and for which no clear-cut ruling could be found in the Qur'an or Prophetic precedent; how to apply the principles of inheritance outlined in the Qur'an to complex situations that had not existed in the life of the Muslim community in Madinah. These and numerous other issues arising out of a rapidly expanding social, political, and economic landscape produced a fervor of intellectual activity that resulted in the emergence of new branches of knowledge.

The intellectual revolution that took place in the world of Islam at this early stage was as much a result of the internal dynamics of the unfolding of Islam in history as it was due to its encounter with some of the richest intellectual traditions of the ancient world. Already during the life of the Prophet, the sciences of the Qur'an (*Ulum al-Qur'an*) had emerged as a differentiated branch of learning. Shortly after his death, focused efforts were made to preserve, annotate, and verify the *Hadith* (narrated sayings of the Prophet).

The Qur'an consists of about 70,000 words, and, for all practical purposes, its text has remained without dispute through history. It existed in written form during the life of the Prophet and was compiled shortly after his death in the form of a book by Abu Bakr, his first successor (Azami 2003). The *Hadiths*, however, were in the thousands, and only a portion of these sayings of the Prophet had existed in written form during his lifetime. His Companions started to compile these sayings after his death, which led to the emergence of a rich tradition of *Hadith* studies, which in turn called for the development of exacting methodologies and techniques to authenticate, index, and cross-reference a very large number of individual sayings of the Prophet. This collection activity, which received sustained and focused attention by successive generations of Muslims until about the middle of the third century of Islam, produced verification methodologies that were later employed by scientists and philosophers in other fields, such as the exact sciences.

The Qur'an and the *Sunnah*—the combined body of the Prophet's sayings and his traditions—are the two primary sources of Islam, and both have remained at the heart of Islamic tradition for over fourteen hundred years. For the purpose of this book, it is important to mention that various branches of learning dealing with these two primary sources emerged in Islamic civilization prior to any other branch of knowledge, and they influenced all other fields—including the natural sciences—in numerous direct and indirect ways. Thus, by the time the study of nature appeared in Islamic civilization as an organized and recognizable enterprise, the religious sciences had already been firmly established; this sequence affected the framework used to explore nature.

The Qur'an itself lays out a well-defined and comprehensive concept of the natural world, and this played a foundational role in the making of the scientific tradition in Islamic civilization. It is therefore incumbent to briefly mention the Qur'anic view of nature in order to develop a methodology for exploring specific aspects of the relationship between Islam and science.

THE QUR'AN AND THE NATURAL WORLD

Although the three Abrahamic monotheistic religions—Judaism, Christianity, and Islam—share a certain degree of commonality with respect to belief in one God and certain aspects of creation, their concepts of the

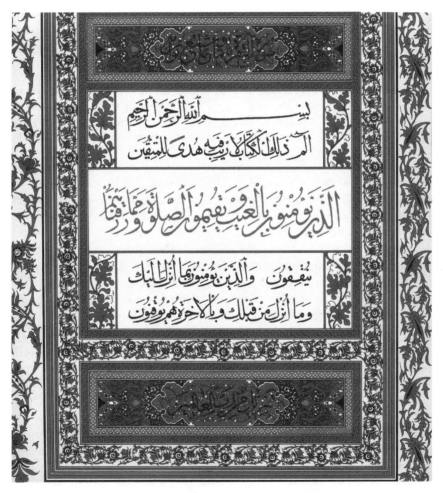

An illustrated page from a thirteenth-century Qur'an manuscript in Thulth and Rayhan scripts. The Qur'an was revealed to Prophet Muhammad over a period of twenty-three years (610–632). The Qur'anic view of nature influenced the development of science in Islamic civilization. © Al-Qalam Publishing.

natural world and its relationship to humanity has considerable divergences (Nasr 1996, 53). In the case of Islam, the Qur'anic view of nature is characterized by an ontological and morphological continuity with the very concept of God—a linkage that imparts a certain degree of sacredness to the world of nature by making it a Sign (*ayah*, pl. *ayat*) pointing to a transcendental reality. However, just as the Qur'an presents the world of nature to humanity as a sign, it also calls its own verses *ayat*. This semantic linkage is further strengthened through various Qur'anic descriptions and modes of communication. God communicates, according to the Qur'an, by "sending" His *ayat*. As Izutsu has noted,

> on this basic level, there is no essential difference between linguistic and nonlinguistic Signs; both types are equally divine *ayat* . . . the meaning of this, in the sense in which the Qur'an understands it, is that all that we usually call natural phenomena, such as rain, wind, the structure of the heaven and the earth, alternation of day and night, the turning about of the winds, *etc.*, all these would be understood not as simple natural phenomena, but as so many 'signs' or 'symbols' pointing to the Divine intervention in human affairs, as evidences of the Divine Providence, care and wisdom displayed by God for the good of human beings on this earth. (Izutsu 1964, 142–143)

Thus science, as a systematic study of nature and as it developed in Islamic civilization, could not treat nature and its study as an entity *separate* from Islam. Furthermore, the Qur'an views nature as a vast system pregnant with movement rather than an inert body. Nature accepts and acts upon Divine Commands, like all else between the heavens and the earth. This view of nature grants it distinct metaphysical qualities. Rather than being self-subsisting, autonomous, or random, nature is described by the Qur'an as a sophisticated system of interconnected, consistent, uniform, and highly active entities, all of which are ontologically dependent on the Creator and exalt Him in their own specific ways (Q. 24:41). *The seven heavens and the earth and whatever is between them sing the glories of God*, an oft-repeated refrain of the Qur'an tells us. It must be noted here, however, that this dependence and subservience of nature to God is not a haphazard matter, since God's ways and laws are unchanging (Q. 33:62), and thus the entire world of nature operates through immutable laws that can be discovered through the investigation of nature. Since these laws are both uniform and knowable, and since nature points to something higher than itself—indeed, to the Creator Himself—it follows that the study of nature leads to an understanding of God, and is in fact a form of worship.

In understanding these relationships drawn by the Qur'an, it is important to recall that the Qur'an is considered by Muslims to be the actual speech of God, imparted to the heart of the Prophet by the Archangel Gabriel. The Prophet then conveyed it as he received it. The text of the

Qur'an thus becomes the actual Divine Word, not retrojective inspirational transcriptions, and so its conception of the natural world is (for the Muslim) grounded in immutable faith.

It should also be noted that, according to the Qur'an,

human beings were created by God as His vicegerents (*khalifa*) in the physical world lying within the finite boundaries of time ... [and] the very principle of God's vicegerency also made them His servants (*'abd, 'ibad*) who were—by virtue of a Primordial Covenant (*mithaq*) they had affirmed, and a Trust (*amanah*) they had taken upon themselves in preeternity—the custodians of the entire natural world. Humanity was thus transcendentally charged not to violate the 'due measure' (*qadr*) and balance (*mizan*) that God had created in the larger cosmic whole. (Haq 2001, 112)

The Qur'anic view of the relationship between the world of nature and God on the one hand and between the world of nature and the progeny of Adam on the other is thus highly interconnected.

Adam's superiority over other creatures and his regency over nature arise in a context that is highly complex, with its interdigitating metaphysical, moral, and naturalistic dimensions: the conceptual setting here evidently being very different from that of the Old Testament and the Evangel. (Haq 2001, 112)

Another aspect of the Qur'anic view of nature is intimately linked to the Divine Name *al-Rahman*, the Most Merciful (one of the two names of mercy to be found in the *Basmallah*, the verse of the Qur'an placed at the head of all but one of its chapters: *In the name of Allah, the Most Merciful, the Most Beneficent*). It has been frequently noted that, in the totality of the Qur'anic teachings, God's Mercy and His Omnipotence are inseparable (Haq 2001, 118), and that "these two perfections are the two poles of divine action, at the same time contrasted and complementary" (Haq 2001). According to the Qur'an, the very act of bringing into existence from nonexistence is an act of mercy. Furthermore, the Qur'anic view of existence not only involves this first act of mercy but also the idea that the continuous existence of things is entirely due to their being sustained by God, one of Whose Names is *Rabb*, the One Who nourishes and sustains. Thus, according to the Qur'an, not only the act of creation but also the act of providing sustenance for the continuation of existence is also an act of mercy—and, since nature is an expression of God's Mercy, all that it contains is by its sheer bountiful existence an undeniable sign of *God's* existence (and Mercy).

The theme of Mercy is especially relevant to our subject. In a chapter of the Qur'an entitled *al-Rahman*, a vast range of phenomena of the natural world—the sun and the moon, rivers, oceans, fruits, cattle—are mentioned as being so many divine favors and blessings. The Qur'an then asks, in a powerful refrain occurring thirty-three times in this chapter, *which of the favors of your Sustainer will you then deny?*

One can say that the whole thrust of the Qur'anic view of nature and its relationship with God and humanity is underscored by three interconnected concepts: *Tawhid* (Oneness of God) and the various associated concepts arising out of the manifestation of divine attributes; the *Amr* (Command) of God in the operational realm; and the intertwined presence of *Qadr* (Measure) and *Mizan* (Balance) in the material world (Haq 2001). These three Qur'anic concepts are not only central to the teachings of Islam but they are also of immense importance for understanding the relationship between Islam and science. Although Islam, "like Judaism, Christianity, Hinduism, and many other religions, has developed numerous schools of thought [—] theological, philosophical, scientific, and mystical [—] dealing with the order of nature" (Nasr 1996, 60), these three concepts are central to the understanding of nature in all schools of Islamic thought.

Given these inherent relationships between God, humanity, and nature, it is impossible in Islam to conceive of nature as an independent, self-subsisting entity. Likewise, science—as an organized enterprise that studies and explores the natural world—cannot be conceived as a separate entity which has to be somehow externally related to Islam. In fact, the much-touted lack of separation of state and religion in Islamic polity is applicable to all other domains, as Muslims believe that Islam is not merely a set of commandments and rituals but a complete way of life, encompassing all domains of knowledge and human activity. This worldview is based on an uncompromising insistence on *Tawhid*, the Oneness of God, a ubiquitous concept in Islamic thought that unifies all realms of knowledge, making them branches of the same tree. Difficult as it may be for the modern Western mind—accustomed to regarding religion solely as set of personal beliefs—to understand this aspect of Islam, it is impossible to construct a relationship between Islam and science—or any other domain of knowledge—as a relationship between two distinctively separate entities.

We need to understand this relationship like that of a mother and a child, in which a particular branch of knowledge—science—emerges from within the greater body of knowledge dealing with the world of existing things, a world conceived as created by and ontologically dependent upon the Creator. It is a relationship that is inherently inseparable from the well-articulated concept of nature as a Divine Sign.

The next chapter explores the emergence of science in Islamic civilization, its relationship with the Greek, Persian, and Indian scientific traditions and its flowering.

Chapter 2

Aspects of Islamic Scientific Tradition (the eighth to the sixteenth centuries)

THE EMERGENCE OF SCIENCE IN ISLAMIC CIVILIZATION

Our current knowledge of original sources does not permit us to draw a clear picture of the initial phases of scientific enterprise in Islamic civilization. What can be said with a certain degree of confidence pertains to the period beginning with the second half of the second century of Islam. By then, however, the enterprise of science in Islam was already well established, with definable branches and scientists of high caliber working in disciplines such as cosmology, geography, astronomy, and alchemy. Thus, until we discover new manuscripts and other primary sources, the story of the emergence of science in Islamic civilization has to remain tentatively dated around 777, the year in which Jabir bin Hayyan, one of the most accomplished Muslim scientists of this early period, is said to have died.

Despite the paucity of early sources, we can confidently trace two branches of science—medicine and astronomy—to the days of the Prophet himself, because we do have verifiable sources allowing us to recount the story of their emergence in Madinah. Sayings of the Prophet dealing with health, sickness, hygiene, and specific diseases and their cures were compiled and systematized by later generations of Muslims, and this body of literature provided the foundation for a specific branch of medical science in Islam: *al-Tibb al-Nabawi*, Prophetic Medicine. Numerous books on Prophetic medicine have preserved for us not only early accounts of how this branch of medicine emerged but also sophisticated theoretical discussions on the entire range of subjects dealing with health and medicine in Islam (al-Jawziyya 1998). Likewise, pre-Islamic Arabic astronomy was radically transformed under the influence of Qur'anic cosmological doctrines to give birth to characteristically Islamic astronomical literature generally referred to as the radiant cosmology (*al-hay'a as-saniya*).

These early sciences had practical use for the first community of Muslims living in Madinah in the new Islamic state, but it is not merely their utilitarian aspect that is of interest to us here; what concerns us at the outset are the intrinsic links of these sciences with Islam. The very foundations of these two branches of science can be shown to have direct connections with the Qur'an and *Sunnah*, the two sources that define all things Islamic. "It is not a coincidence," notes George Saliba, "that the mathematical astronomical tradition which dealt with the theoretical foundations of astronomy also defined itself as a *hay'a* [cosmological] tradition, even though it rarely touched upon the Quranic references to the cosmological doctrines" (Saliba 1994, 17). Likewise, other sciences that emerged in Islamic civilization can be shown to have intrinsic links with the Islamic worldview, even though they received a large amount of material from other civilizations. These links and connections will remain our continuous focus as we construct our narrative about the emergence of science in Islamic civilization.

The geographical expansion of Islam within its first century was, as noted earlier, accompanied by a social revolution that reconfigured the social, cultural, and intellectual climate of the old world. The same social revolution provided an opportunity for Islamic civilization to receive a very large amount of scientific material from Greek, Persian, and Indian sources. This infusion was not a random process; rather, it was an organized and sustained effort spread over three centuries, involving thousands of scientists, scholars, translators, patrons, books, instruments, and rare manuscripts. But it must be pointed out as we approach this fascinating tale that this process could not have occurred without the ability of the recipient civilization to absorb. In other words, prior to the arrival of Greek, Indian, and Persian scientific material, there must have been an indigenous scientific tradition ready and able to comprehend and receive this material. We know, for instance, that as early as the second quarter of the eighth century, astronomical treatises were being written in Sind (modern Pakistan) in Arabic. These early treatises were often based on Indian and Persian sources, but they employed technical Arabic terminology that could not have come into existence without the presence of an already-established astronomical tradition in Islamic civilization that could then absorb new material from Indian and Persian sources.

As we proceed with the account of the emergence of science in Islamic civilization, we should note that the Islamic scientific tradition was emerging in a cosmopolitan intellectual milieu and that those who were making this tradition were not only Muslims but also Jews, Christians, Hindus, Zoroastrians, and members of other faith communities. An Indian astronomer who arrived in the court of the Abbasid Caliph al-Mansur (r. 754–775) as part of a delegation from Sind, for instance, was probably a Hindu. He knew Sanskrit and helped al-Fazari (*fl.* second half of the eighth century) translate a Sanskrit astronomical text into Arabic; this text contained elements from even older astronomical traditions. The resultant translation, *Zij al-Sindhind*, became one of the sources of a long tradition of such texts in Islamic astronomy (Pingree 1970, 103–23).

The emergence of the scientific tradition in this multireligious, multiethnic atmosphere was a dynamic process involving interactions between patrons of learning, scholars, scientists, rulers, guilds, and wealthy merchants. To be sure, the scientific activity at this early time was not yet an institutionalized effort, but we do know that groups of scientists were already working together in the second half of the eighth century in Baghdad and other cities of the Abbasid caliphate. We should also keep in mind that this scientific tradition was evolving at a time when the religious sciences had already been established on a firm foundation, with advanced texts in Qur'anic studies, philology, grammar, jurisprudence, and other branches of religious studies circulating among scholars. This fact is particularly important for our study because the prior establishment of religious sciences meant that the new scientific tradition emerged into an intellectual milieu already shaped by religious thought.

In the atmosphere of intense creativity that permeated Islamic civilization during this early period there was considerable strife and polarization at all levels of society. By the time science emerged as a differentiated field of study, Islamic polity had already gone through two major internal fissures: the first (656–661) was sparked by the assassination of Uthman, the third Caliph, and led to a civil war in which close Companions of the Prophet found themselves pitched against each other under circumstances which threatened the very existence of the Muslim community. During the second rift (680–692), which sprang from two rival claims to the Caliphate, Husayn bin Ali, the grandson of the Prophet, and all but one of his companions were killed at Karbala in October 680; Makkah was besieged by armed men; a radical splinter group, the Khawarij, took control of much of Arabia; and Ibn al-Zubayr, one of the close Companions of the Prophet, was killed in the sanctified city of Makkah, where fighting had been declared unlawful by the Qur'an (Q. 2:217).

These events initiated an intense debate among scholars, not only about *what* was happening and *why* but also certain other fundamental issues that arose in this context: Is this a crisis of leadership? Who is qualified to lead the community? Are human acts preordained? What are the boundaries of human freedom? What is the role of human intellect in matters of religion? What is the exact nature of Divine justice, Hell and Heaven, and that of Divine attributes? These and related theological debates eventually gave birth to different schools of thought; some of these schools also developed their own positions on the natural sciences, and we will have occasion to discuss their positions in a later chapter.

The period during which the earliest scientific works were written witnessed a revolt against the Umayyads, who had taken control of the Caliphate and shifted the capital of the Islamic state from Madinah to Damascus. Originating in newly conquered Iranian cities, especially in Merv, this revolt in favor of the Abbasid claim to the Caliphate moved westward under the leadership of Abu Muslim, who had captured the city of Kufa by the middle of 749. Early in 750, Abu'l Abbas (posthumously called al-Saffah)) was proclaimed the first Abbasid Caliph at Kufa. Two months later, the Umayyads were decisively defeated at the battle of Greater Zab, and by June 750 most of them had been massacred. Abd al-Rahman I was the sole survivor from the ruling family; he escaped to Spain, where he established Umayyad rule (755–1031).

Abu'l Abbas remained at war for the entire period of his caliphate and on his death in 753, his brother, Abu Ja'far, was proclaimed Caliph as al-Mansur (the victorious). In 762, al-Mansur decided to move his capital to a safer place. He himself supervised the process of the selection of its location; the choice fell for a small and ancient town, which was to become the fabled capital of Abbasid rule for the next five hundred years: Baghdad. Spanning both banks of the river Tigris, the new capital was designed as a circular city with sixteen gates. Its construction began on July 30, 762, a date determined by astrologers and engineers, among whom was the aforementioned al-Fazari. The city was officially called *Madinatul-Islam*, the city of peace.

Beginning with the construction of Baghdad we can trace the developments in the scientific tradition in Islamic civilization with more confidence; our source material becomes more reliable and there is an exponential increase in available texts.

In order to understand the nature of science in Islamic civilization at this stage of its development, we proceed with an outline of the various sciences as they emerged during the second half of the eighth century.

THE INITIAL FLOWERING

By the time of the famous alchemist Jabir bin Hayyan, science in Islamic civilization had become considerably well established. Jabirian corpus is so extensive and varied that some scholars have expressed doubts about its authorship by a single person (Kraus 1991). These highly sophisticated works, dealing with a vast range of subjects, were to leave a legacy that continued to influence science and discourse on science well into the fifteenth

The opening folio of a fragment of an alchemical treatise attributed to Abu Bakr Muhammad ibn Zakariya al-Razi (d. 925) that is otherwise known only through its Latin translation *Liber 70 praeceptorum*. Courtesy of the National Library of Medicine.

Another folio of a fragment of an alchemical treatise attributed to Abu Bakr Muhammad ibn Zakariya al-Razi (d. 925). Courtesy of the National Library of Medicine.

century. Jabir's writings deal with the theory and practice of chemical processes and procedures, classification of substances, astrology, cosmology, theurgy, medicine, alchemy, music, magic, pharmacology, and several other disciplines. What provides an internal cohesion to this corpus is the overall framework of inquiry and, more specifically, his "Theory of Balance." According to Jabir, all that exists in the cosmos has a cosmic balance. This balance is present at various levels and reflects the overall harmony of all that exists.

In addition to Jabir, many lesser-known scientists of this period demonstrated keen interest in astronomy, mathematics, cosmology, and medicine. Only a small number of fragmented works from this period have so far been studied, and this does not allow us to traverse the early history of Islamic scientific tradition. Texts available to historians of science take us directly into the first half of the ninth century, when Baghdad had already become the intellectual and scientific capital of the Abbasid empire, providing scientists patronage and opportunities to experiment, discuss, and discover. Most of these scientists were interested in more than one branch of science, as was usual at that time. The highest concentration of scientific activity at this early stage is, however, in mathematics, astronomy, alchemy, natural history, and medicine.

It is important to pay attention to this early period of Islamic scientific tradition, because the massive amount of Greek works subsequently translated into Arabic have created the erroneous impression that Islamic scientific tradition came into existence through the Translation Movement, and that all it did was to preserve Greek science for later transmission to Europe.

GREEK CONNECTION

That the Islamic scientific tradition preceded the translation movement, which brought a large number of foreign scientific texts into this emerging tradition, is beyond doubt; even our meager resources amply prove this. Astronomy, alchemy, medicine, and mathematics were already established fields of study before any major translations were made from Greek, Persian, or Indian sources. Translations were done to enrich the tradition, not to create it, as some Orientalists have claimed.

In the field of astronomy, for example, George Saliba (1994), E. S. Kennedy (1967), and David Pingree (1970) have conclusively shown that a very accomplished generation of astronomers, which included Yaqub b. Tariq (*fl.* second half of the eighth century) and several others, was already at work before the great translation movement (Saliba 1994, 16). Saliba also shows how this early astronomical tradition was related to Qur'anic cosmology. Pre-Islamic astronomy (known as *anwa'*), which predicted and explained seasonal changes based on the rising and setting of fixed stars, was a subject of interest for Qur'anic scholars as well as for the early lexicographers, who produced extensive literature on the *anwa'* and *manazil* (lunar mansions) concepts (Saliba 1994, 17).

The large amount of scientific data and theories that came into the Islamic scientific tradition from Greek, Persian, and Indian sources were not simply passively translated for later transmission to Europe. In fact, translated material went through constant and detailed examination and verification, and was accepted or rejected on the basis of experimental tests and observations. This process of scrutiny started as early as the ninth century—that is to say, almost contemporaneously with the translation movement. The tradition of the production of astronomical tables, for instance, may have been inspired by the Ptolemaic *Handy Tables* tradition, but the tables produced by Muslim astronomers were not merely an Arabic reproduction of Ptolemaic tables; they were the result of astronomical observations that began as early as the first half of the ninth century with the expressed purpose of "updating the *Zijes*, inspired by the *Handy Tables*" (Saliba 1994, 18). Furthermore, as Saliba points out,

no astronomer working in the early part of the ninth century could still accept the Ptolemaic value for precession, solar apogee, solar equation, or the inclination of the ecliptic. The variations were so obvious that they must have become intolerable and could no longer be explained without full recourse to a long process of questioning the very foundation of the validity of all the precepts of Greek astronomy. (Saliba 1994, 18)

This critical attitude toward received material was neither accidental nor a passing phenomenon; among other things, it gave rise to a novel tradition of *shukuk* literature, which cast doubts on various theoretical assumptions of Greek science, called for a reexamination of observational data, produced texts that dealt with internal contradictions in each branch of Greek science, and produced a critical attitude toward the translated texts, which spurred a movement for their revision, both at the level of experiment and theory. Abu Bakr Zakaria al-Razi's yet-to-be-published *Kitab al-Shukuk Ala Jalinus* (*Doubts Concerning Galen*), and Ibn al-Haytham's *al-Shukuk Ala Batlamyus, Dubitationes in Ptolemaeum* are excellent examples of this kind of literature (Sabra and Shehaby 1971). The translation movement is examined in more detail in a subsequent section of this chapter.

ISLAM AND ITS SCIENTIFIC TRADITION

Was there any connection between Islam and the scientific tradition that was emerging in Islamic civilization in the eighth century? Can this science be called "Islamic science"? These two questions are central to this book and will be examined throughout, but it may be beneficial to briefly mention the current prevalent position in this regard, which holds that Islam had nothing to do with the scientific tradition that emerged in the Islamic civilization. In fact, this approach is not specific to Islam; such accounts

Opening page of Ibn al-Haytham's *Kitab al-Manazir (Optics)* from an old manuscript. Courtesy of Maktabah al-Fatih. © Maktabah al-Fatih.

of science conceive all sciences, at all times and all civilizations, to be enterprises totally independent of all religions—and if any interaction between religion and science becomes unavoidable, it is normally perceived as negative. For numerous reasons, this opinion regards any relationship between Islam and science with extra suspicion. Some even go as far as to say there is, in fact, no such thing as a normative Islam, and that all we can say with certainty is that there are numerous kinds of Islam—an

Islam of the Makkan period, an Islam of the time when the Prophet was establishing a state in Madinah, an Umayyad Islam, an Abbasid Islam, and so on (Gutas 2003). This approach to the question of Islam's relationship with science not only rejects the notion of anything that can be called "normative" or "essential" Islam, it also claims that

> Islam, as a religion, and at whatever historical moment it is taken, is a specific ideology of a particular, historically determined society. As such, like all other social ideologies that command adherence and respect by the majority of the population because of their emotive content, it is inert in itself and has no historical agency but depends completely on who is using it and to what ends. (Gutas 2003, 218)

Gutas is not alone; battalions of latter-day postmodernists, secular historians of science, neo-Orientalists, and even sociologists who have an aversion to religion hold the same view under the influence of contemporary postmodernism. This comes in stark contrast to the nineteenth- and early-twentieth-century Orientalists, who spent all their energies in constructing a homogeneous Islam in which an "orthodoxy" could be identified and posited against an opposing tradition of "free thinking." Since the last decade of the twentieth century, and more so since the beginning of the twenty-first century, the various effects of postmodernism have been busy at deconstruction and the creation instead of a fluid Islam that has nothing stable at any level. Thus, instead of the monolithic, homogenized, rarified, and static Islam of the Orientalists, we now have an Islam that can be fundamentally different across—and within—regions and eras. Needless to say, both extremes have added little clarity to the conceptual categories so essential for real communication.

Here we are brought to an interesting contradiction in much of this thought: even though it is claimed there is no "essential Islam," one can still safely speak of some "Islamic" phenomena—for example, Islamic calligraphy and Islamic poetry. While the possibility of an "Islamic science" is immediately denied, the "Islamic garden" and "Islamic architecture" do not undergo the same vehement reductionism. Furthermore, and even more interesting is that while denying Islam any essential nature, proponents of this thought create an essential *science* separate from any wider context or framework.

Such accounts of the scientific activity in Islamic civilization ignore the Qur'anic conception of nature outlined through many verses, giving us a systematic and coherent view of the subject of scientific investigation—nature. Because of the antagonism toward the foundational relationship between Islam and the scientific tradition that was cultivated in Islamic civilization for eight hundred years, such accounts also fail to adequately explain the development of those branches of science that were directly related to Islamic practices: astronomy used to determine the distance and direction toward Makkah (the direction Muslims face for their obligatory prayers five times a day); geography; geodesy; cartography; *mawaqeet* (the science of timekeeping); and other such branches of science that have a direct relationship with Islamic practices. These are not simply the cases of "science in the service of religion," as is sometimes claimed; rather, these sciences emerged from a specific view of nature anchored in Islam.

The contemporary quasi-postmodern approach to Islam has also created an academic atmosphere, which inhibits empirical studies of the connections between Islam and the scientific tradition that existed in Islamic civilization prior to the modern era. When seen in its proper perspective, Islam is not a fluid conceptual framework that keeps changing with time; rather, an Islamic way of being can be verifiably traced back to the life of the Prophet of Islam—a life lived in the full light of history and preserved with great care for posterity. This concrete and real life of Muhammad is at the heart of the Islamic way of life. This life, which is considered to be a living model of the Qur'an, is not an abstract idea needing theological interpretation. Thus, while it is true that within the broad contours of the Islamic civilization all kinds of rulers, patrons of learning, scholars, and scientists have existed and continue to exist, and that what any individual ruler believed or believes may influence the course of Islamic civilization to some extent, no individual *defines* it. Islamic civilization is, as any other civilization, defined by its belief system, a priori presuppositions, and a legal and moral framework. It is this framework arising out of Islamic beliefs and practices that created the matrix from which intrinsic links between religion and the sciences grew and flourished in the Muslim lands.

Another dimension of these studies has to do with hasty judgments passed regarding the overall achievements of Islamic scientific tradition and with setting its demise in the twelfth century. Both of these judgments were passed early in the nineteenth century, when only a fraction of the source material available today had been discovered and studied, but they continue to remain the mainstream version. David King has recently lamented in his monumental work *In Synchrony with the Heavens: Studies in Astronomical Timekeeping and Instrumentation in Medieval Islamic Civilization*:

Some out-dated notions wide-spread amongst the "informed public" and even amongst historians of science are that:

(1) The Muslims were fortunate enough to be the heirs to the sciences of Antiquity.
(2) They cultivated these sciences for a few centuries but never really achieved much that was original.
(3) They provided, mainly in Islamic Spain, a milieu in which eager Europeans emerging out of the Dark Ages could benefit from these Ancient Greek sciences once they had learned how to translate them from Arabic into Latin.

> Islamic science, therefore, one might argue, is of no consequence *per se* for the development of global science and is important only insofar as it marks a rather obscure interlude between a more sophisticated Antiquity and a Europe that later became more civilized.
>
> What happened in fact was something rather different. The Muslims did indeed inherit the sciences of Greek, Indian and Persian Antiquity. But within a few decades they had created out of this potpourri a new Muslim contributions, which flourished with innovations until the 15th century and continued thereafter without any further innovations of consequence until the 19th. (King 2004, xvi)

Despite the large amount of new material discovered, published, and studied since those early notions were formed, not many contemporary writers are willing to reexamine the erroneous paradigm postulated by Goldziher and his generation, which pit Islam against "foreign sciences" (Goldziher 1915). These early judgments were also based, in part, on the works of medieval European scholars who themselves were aware of only a miniscule body of literature on Islamic scientific tradition, mostly retrieved from Islamic Spain (al-Andalus), a region that lay outside the main centers of Islamic scientific activity. It was not until the nineteenth and twentieth centuries

> when historians of science from a multiplicity of national backgrounds investigated Islamic scientific manuscripts in libraries all over Europe and then in the Near East. Their investigations revealed an intellectual tradition of proportions that no medieval or Renaissance European could ever have imagined: anyone who might doubt this should look at the monumental bio-bibliographical writings of Heinrich Suter, Carl Brokelmann, and Fuat Sezgin. (King 2004, xvii)

Even though King's book is concerned with only one aspect of Islamic science (astronomical timekeeping and instrumentation), it has brought to the field of history of Islamic science a large amount of new material

> which has become known only in the past 30 years. Inevitably [it] modifies the overall picture we have of Islamic science. And it so happens that the particular intellectual activity that inspired these materials is related to the religious obligation to pray at specific times. The material presented here makes nonsense of the popular modern notion that religion inevitably impedes scientific progress, for in this case, the requirements of the former actually inspired the progress of the latter for centuries. (King 2004, xvii)

Since this book is not on the history of Islamic scientific tradition but on the relationship between Islam and science, it cannot go into further details, but it is clear that what remains to be recovered and studied from the original sources in various branches of science is far greater than what has been studied so far, and that a final assessment of the Islamic scientific tradition can only be made after further source material has been carefully examined by competent historians and scholars.

Before exploring various aspects of the Islam and science relationship, it must be pointed out that sciences cultivated in Islamic civilization were not always the work of Muslims; in fact, a considerable number of non-Muslims were part of this scientific tradition. What made this science Islamic were its integral connections with the Islamic worldview, the specific concept of nature provided by the Qur'an, and the numerous abiding concerns of Islamic tradition that played a significant role in the making of the Islamic scientific enterprise. There were, of course, at times bitter disputes between proponents of various views on the nature of the cosmos, its origin, and its composition, but all of these tensions were *within* the broader doctrines of Islam, which conceived the universe in its own specific manner—a definable, specific, and distinct conception that placed a unique, personal, and singular Creator at the center of all phenomena. Viewed from this perspective, the Islam and science nexus can be explored as a much more fruitful encounter within the greater matrix of Islamic civilization.

As already mentioned, our current knowledge of primary sources about the first half of the eighth century does not permit us to trace the beginning of the natural sciences in Islamic civilization in detail. By the end of that formative century, however, there was already a small and vibrant scientific community whose members were exploring the world of nature in a milieu filled with intellectual curiosity and creative energy. As was usual at that time, this community consisted of individuals who were interested in a wide range of subjects dealing with nature, history, and philosophy, and not with just one particular branch of science.

ISLAMIC SCIENCE OR NATURAL PHILOSOPHY?

Their work is sometimes called *natural philosophy* rather than *science*. This term is also used for the enterprise of science that existed in the Greek and Roman civilizations. This linkage adds weight to the view that science in Islamic civilization was somehow merely an extension of the earlier Greek and Roman science. There was, however, no one term in Greek or Latin equivalent to our contemporary term *science*, and what we understand as *science* was often called *philosophy* or *inquiry concerning nature* by the Greeks and Romans themselves (Lloyd 1973, xi–xiii). Unlike Greek and Latin, Arabic does have a specific word for science: *al-ilm*. This word as well as its derivatives frequently occurs in the Qur'an. It is used to denote all kinds of knowledge, not just the knowledge pertaining to the study of nature, but this semantic linkage of all branches of knowledge does not mean that knowledge was not differentiated or classified into various hierarchical branches. Rather, it indicates that within a given classification of knowledge, all branches of knowledge were intimately linked through a vertical axis running through the entire epistemological scheme—a grounding in the Qur'anic concept of knowledge.

It is, therefore, conceptually problematic to use the Aristotelian term *natural philosophy* as an equivalent for those branches of knowledge that dealt with the study of nature in Islamic civilization. This term may be a correct way of describing Greek and Roman scientific traditions, but its use here is applied to a very different conceptual scheme. Although a large amount of scientific data from the Greek tradition came into Arabic, this transfer was not accompanied by an incorporation of the Greek epistemology from which the term *natural philosophy* originally emerged.

The term *natural philosophy*, often used interchangeably with *physics*, emerged from within the Aristotelian classification of knowledge into three broad categories: metaphysics, mathematics, and physics. Metaphysics deals with unchanging things such as God and spiritual substances; mathematics studies unchanging abstractions not God or spiritual substances; and physics studies changeable things in the natural world, including both animate and inanimate bodies. With regard to physics, although he accorded a high degree of importance to sense perception, he maintained that knowledge about nature cannot be derived by means of sense perception alone; to attain scientific knowledge about the physical world, universal propositions—obtained from sense perceptions by means of induction—are essential (Aristotle 1984, 132).

Aristotle's entire classification scheme, however, is ultimately dependent on his idea of God and the creation of the world. He believed in an eternal world, ultimately caused by an impersonal deity eternally absorbed in self-contemplation. His eternal world was rationally structured and comprehensive, but it was, nevertheless, a world without any direct involvement of the deity. It was a world in which bodies were composed of matter and form and in which change was caused by four types of causes: material, formal, efficient, and final. These causes could produce four kinds of changes in a body: substantial change involved change of form (wood to ashes by fire); qualitative change involved change of a certain quality of body (the change of a green leaf into a yellow leaf); quantitative change involved change in the size of a body; and change of place involved movement of the body from one area to another. Thus, for Aristotle, the study of nature by means of natural philosophy was the study and analysis of causes and the changes these causes produced. His natural philosophy (physics) embraced all bodies, and included the study of the processes of generation and corruption of compounded bodies from four simple substances (earth, fire, air, water) as well as the study of animals and plants. Since in Aristotle's conception of the domain of knowledge, *natural philosophy* and *physics* are synonymous, the same terminology is sometimes applied to the Islamic scientific tradition, where the Arabic word *Tabi'at* (physics) is used to describe all branches of science as well as physics itself. This usage can lead to complications, especially when individual branches of science are known to have their own specific names.

Aristotle's concept of God as well as his belief in the eternity of the world was in direct opposition to the Qur'anic concept of God and the world. Thus, even though a large amount of Aristotelian philosophy was incorporated into the Islamic philosophical tradition, his deity was unacceptable. The translations of his works thus created a tension within the emerging Islamic scientific tradition. The subsequent story of the interaction of Islam and science is, to a large extent, a story of how Muslim philosophers and scientists dealt with this tension. We will explore various facets of this tension in the next section.

When the study of nature emerged in Islamic tradition as a fully differentiated field, it found its place within a preexisting framework of classification of knowledge. This classification scheme follows a certain pattern based on the Qur'anic concepts both of knowledge and the faculties granted to human beings. Within this study of nature, innumerable specific disciplines emerged, which were in turn refined and further distinguished. Thus, for example, we have titles like the celebrated *Kitab al-Manazir* (*Optics*) of Ibn al-Haytham and even more specific titles like *Kitab tahdid nihayat al-amakin li'tashih musafat al-masakin* (*The Book for the Determination of the Coordinates of Positions for the Correction of Distances between Cities*) of al-Biruni. It was also common to use titles such as *Kitab al-Nujum* (*Book on Stars*) for works on astronomy and *Kitab ilm al-hindasah* (*Book on the Science of Geometry*) for works on geometry. Certain Muslim philosophers more heavily influenced by Aristotle (e.g. al-Kindi, d. *ca.* 873; al-Farabi, d. 950; Ibn Sina, d. 1037; and Ibn Rushd, d. 1198) did in fact utilize the Aristotelian model for classification of knowledge, but even they had to modify his essential elements in order to incorporate the basic belief system of Islam. Thus even those schemes of classification of knowledge that were heavily influenced by Aristotle retained essential Islamic concepts regarding God, human beings, and the nature of this world. Other classification schemes, especially those of al-Ghazali (d. 1111) and Ibn Khaldun (d. 1406), attempted to remove Aristotelian influences altogether.

THE TRANSLATION MOVEMENT AND ITS IMPACT ON THE DEVELOPMENT OF SCIENCE IN ISLAMIC CIVILIZATION

From about the middle of the eighth to the middle of the eleventh century, a systematic, elaborate, sustained, and well-organized translation movement brought almost all philosophical and scientific books available in the Near East and the Byzantine Empire into Arabic. This translation movement has now been the focus of scholarly studies for over a century and a half, and this scholarship has documented a great deal of historical data and information. Thanks to the discovery and study of numerous manuscripts, we can identify numerous Greek, Pahlavi, and Indian works and their translators, as well as subsequent translators. The scope of this translation movement can be judged from the range of subjects covered, which included the entire Aristotelian philosophy, alchemy, mathematics, astronomy, astrology, geometry, zoology, physics, botany, health sciences, pharmacology, and veterinary science. The extent of social, political, and financial patronage this movement received can be gleaned from the social classes that supported it, and included caliphs, princes, merchants, scholars, scientists, civil servants, and military leaders.

Over the past 150 years, the study of this translation movement has yielded many valuable texts that have enhanced our understanding of the role of Greek science and philosophy in the making of the Islamic scientific tradition. At the same time, inaccuracies and stereotypes have crept into some of these accounts, and this is especially true for those works that attempt to identify the translation movement as the main cause of the origination of Islamic scientific tradition. Thus, it has been claimed that

> the translation movement was the result of the scholarly zeal of a few Syriac-speaking Christians who ... decided to translate certain works out of altruistic motives for the improvement of society. The second theory, rampant in much mainstream historiography, attributes it to the wisdom and open-mindedness of a few "enlightened rulers" who, conceived in a backward projection of European enlightenment ideology, promoted learning for its own sake. (Gutas 1998, 3)

Gutas claims both of these theories fall apart under close scrutiny. He states that the translation movement was "too complex and deep-rooted and too influential in a historical sense for its causes to fall under these categories—even assuming that these categories are at all valid for historical hermeneutics" (Gutas 1998, 4).

This movement was unprecedented in the transmission of knowledge. It was a movement that enriched the Arabic language by forcing its philologists to coin new technical terms, forced the best minds of the time to find ways to accommodate, discard, or transform theories and ideas that conflicted with their religious beliefs, brought a very large amount of scientific and philosophical data into Islamic civilization, and produced tensions and conflicts within the Islamic intellectual tradition that were, in the final analysis, greatly beneficial to the development of Islamic scientific tradition.

Although translation activity had already begun during the pre-Abbasid period, it was the Abbasids who provided resources for a sustained and systematic process of translation of scientific texts into Arabic. The translation movement became more organized and received financial and administrative impetus after the founding of Baghdad by al-Mansur (r. 754–775). Three distinct phases can be identified in this movement. The first began before the middle of the second century of Islam, during the reign of al-Mansur. Major translators of this first phase were Ibn al-Muqaffa (d. 139/756); his son, Ibn Na'ima (*fl.* eighth century); Theodore Abu Qurra (d. *ca.* 826), a disciple of John of Damascus (d. 749) who held a secretarial post under the Umayyad Caliphs; Thabit ibn Qurrah (d. 901), a Sabian mathematician; Eustathius (*fl.* ninth century), who along with Theodore Abu Qurra translated for al-Kindi; and Ibn al-Bitriq (877–944), who was a member of the circle of the Caliph al-Ma'mun. Al-Ma'mun's accession marks the beginning of the second phase of the translation movement.

Many new translators were involved during the second phase of the translation movement, which was led by Hunayn ibn Ishaq. These translators refined many translations of the first phase and extended the range of material being translated. For instance, Aristotle's *Topics* was first translated into Arabic from a Syriac translation around 782, during the reign of the third Abbasid Caliph al-Mahdi (d. 785). This was done by the Nestorian patriarch Timothy I with the help of Abu Nuh, the Christian secretary of the governor of Mosul. The same work was retranslated about a century later, this time from the original Greek, by Abu Uthman al-Dimashqi, and, approximately fifty years later, it was translated a third time by Yahya ibn Adi (d. 974) from a Syriac version (Gutas 1998, 61).

During the third phase of the translation movement, further refinement of the translated material took place and commentaries started to appear. This phase, beginning with the dawn of the tenth century and ending around 1020, also produced textual criticisms that scrutinized translated material from scientific as well as philosophical points of view.

By the middle of the eleventh century the three-hundred-year-old translation movement had reached its end. The tension between Islamic beliefs and ideas, concepts, theories, and data contained in these texts however was the main kinetic force for initiating a process of appropriation and transformation of the received material. This was a slow and deliberate process over the course of which translated works were examined, classified, and sorted into categories. It was not official scrutiny by some office of the state or religious authority, but an organic process of ordering new ideas in the light of revelation undertaken by Muslim intellectuals who debated, disagreed, passed judgments on each other, fought bitter battles over ideas, and lost or gained support from their peers. This inner struggle of a tradition in the making against foreign currents that were coming into its fold involved a wide range of philosophers and scientists. Some of them firmly aligned themselves with the Greek philosophical tradition, while others wrote against it. Those in between these two extremes attempted to harmonize new ideas with Islam's worldview based on revelation. The end result of this long process was the appearance of a tradition of learning that examined, explored, and synthesized its own unique perspectives on nature and the human condition—perspectives that were distinctly Islamic, though not monolithic.

RECENT PERSPECTIVES ON THE TRANSLATION MOMENT

This brief description of the translation movement provides some insights into the intellectual currents flowing into the Islamic tradition during that time. During these three centuries, the material infused into the Islamic tradition included philosophy as well as works on various branches of science drawn from the Greek, Persian, and Indian sources. This process of incorporation of foreign scientific and philosophical thought into the Islamic tradition and its consequences has been thoroughly studied by historians of science and philosophy, and their studies have yielded opinions ranging from *reductionism* to *precursorism*—two explanatory terms first used by Sabra in an important paper (Sabra 1987). Reductionism, in this context, refers to the

view that the achievements of Islamic scientists were merely a reflection—sometimes faded, sometimes bright, or more or less altered—of earlier (mostly Greek) examples. Precursorism (which has a notorious tendency to degenerate into a disease known as 'precursitis') is equally familiar: it reads the future into the past, with a sense of elation. (Sabra 1994, 223–24)

Despite the work of Sabra and a handful of other historians of science, the large-scale infusion of ideas, theories, and scientific data from the Greek scientific tradition into Islamic science through the translation movement has become a defining feature of the Islamic scientific tradition itself; many histories of science tend to regard the eight hundred years of scientific activity in the Muslim world as being no more than some kind of depot where Greek science was parked and from where it was retrieved by Europe in later centuries. As Sabra has noted, the transcivilizational transmission of science was an important event in history, but

apparently because of the importance of that role in world intellectual history many scholars have been led to look at the medieval Islamic period as a period of reception, preservation, and transmission, and this in turn has affected the way in which they have viewed not only individual achievements of that period but the whole of its profile. (Sabra 1994, 225)

The consequences of this approach toward the Islamic scientific tradition are visible in many science textbooks, where students are led to believe that nothing important happened in science between the Greek scientific activity and the Renaissance; Islamic scientific tradition is either not mentioned at all or is mentioned in a paragraph defining it as a repository of Greek science. That this distortion of historical facts still predominates is unfortunate, as what came into the body of Islamic thought from outside was neither accidental nor marginal. It would be far more meaningful to understand

> the transmission of ancient science to Islam . . . as an act of appropriation performed by the so-called receiver. Greek science was not thrust upon Muslim society any more than it was later upon Renaissance Europe. What the Muslims of the eighth and ninth centuries did was to seek out, take hold of and finally make their own a legacy which appeared to them laden with a variety of practical *and* spiritual benefits. And in so doing they succeeded in initiating a new scientific tradition in a new language which was to dominate the intellectual culture of the large part of the world for a long period of time. 'Reception' is, at best, a pale description of that enormously creative act. (Sabra 1994, 226)

The translation movement was a highly complex phenomenon of cross-cultural transmission that involved a very large number of people of diverse interests. It had political, cultural, intellectual, and religious motivations. It came into existence owing to certain internal needs of the Islamic polity of the time and, once in existence, it produced enormous creative energy in an intellectual climate already filled with curiosity, ready to use whatever it could for its new ventures. Some of the new material was regarded as dangerous, extraneous, and foreign by certain quarters. This "foreignness" has been used by some scholars to draw the reductionistic conclusion that the scientific tradition in Islam was nothing but a "foreign" entity that somehow survived despite the opposition it faced from religious circles. This view has been succinctly called "the marginality thesis" and its validity has been challenged on sound historical grounds (Sabra 1994, 229–30). Sabra and other historians of science have also convincingly made a case for the originality of Islamic scientific tradition (Kennedy 1960; Shlomo 1986; King 1999).

Chapter 3

Facets of the Islam and Science Relationship (the eighth to the sixteenth centuries)

To describe the relationship between Islam and the science cultivated in Islamic civilization between the eighth and the sixteenth centuries, this chapter explores specific developments in certain branches of science. This account is, of course, not a comprehensive history of Islamic scientific tradition, but simply an overview of certain developments in those branches of science that had a more direct relationship with religion. Some branches of science (such as mechanics) had little to do with religion in any direct way, whereas others (such as cosmology and geography) had a direct relationship with religion and hence are given more attention in the following sections.

COSMOLOGY, COSMOGONY, AND COSMOGRAPHY

No other branch of science has a more direct relationship with religious beliefs than cosmology—the science that deals with the origin and development of the universe. Yet it is a relationship that is characterized by a great deal of confusion. What is meant by cosmology today is entirely different from what was meant by the same term in the eighth century. The use of similar terms in science, philosophy, and religious discourse has also added to the confusion. For instance, what Aristotle meant by *celestial region* is not at all the *celestial region* of the Sufis, though both may use the term *celestial* to denote the region beyond the terrestrial zone. The celestial region of the Sufis is populated by vastly different entities than that of Aristotle and has totally different characteristics. Cosmology, of course, was more philosophy than science during the period of the Greek and Islamic scientific traditions and even now, while a great deal of experimental data has come into existence that has direct bearing on the question of the origin of the cosmos, it remains a theoretical field.

Islamic cosmological beliefs are rooted in the Qur'an itself, which deals extensively with this issue. Thus, for our purposes, the main question is how these so-called cosmological verses of the Qur'an were understood by exegetes, philosophers, and scientists during the period under consideration (the eighth to sixteenth centuries). Debates arose from the tensions generated by the arrival of Aristotelian cosmology in the Islamic tradition, which, in turn, contributed to the making of certain cosmological doctrines.

As has been stated in the first chapter, the Qur'an treats the entire created order as a sign, *ayah*. This includes the cosmos and all that it contains. A sign, by definition, points to something other than itself. Thus, when seen from the Qur'anic perspective, the cosmos and all that it contains are signs of a unique Creator who created through a simple command: *Be* (Q. 36:82). Although the Qur'an gives a very specific account of the creation of the cosmos, it does not tell us with what it was created or when. In addition, it is important to keep in mind that the Qur'anic cosmos is not merely physical, made up of stars, planets, and other physical entities; it also encompasses a spiritual cosmos populated by nonphysical entities. The nonphysical cosmos, which consists of innumerable levels of existence, is far superior to the physical cosmos, which occupies a relatively low position in the hierarchy of existence.

The Qur'anic perspective on the creation of the physical cosmos can be summarized as follows: the cosmos was created by God for a purpose. After creating the cosmos and all that it contains, God did not leave it to itself; in fact, the entire created order is perpetually sustained by God; without this sustenance it could not exist. At a certain pre-fixed moment, the exact knowledge of which remains with God alone, everything that exists in the world will perish. This will be followed by resurrection and a new kind of life under an entirely new set of laws.

This general account of creation and its end can be found in many verses, supplemented with specific details spread throughout the Qur'an. The cosmos was created in six days (Q. 7:54–56; 25:59), the Earth in two (Q. 41:9); God also created the seven heavens (Q. 2:29), one upon another (Q. 67:3). God adorned the sky with stars (Q. 67:5); He is the One who has set in motion all the stars and planets, so that humanity may be guided in its travels by their positions (Q. 6:97); He is the One Who covers the day with the night and the night with the day (Q. 39:5). It is important to note that the word "day" used in these verses has always been understood in the Islamic tradition in a nonquantitative manner. The Qur'an itself makes it clear that *a day with the Lord is as a thousand years of your reckoning*

(Q. 22:47). In another verse it mentions *a day whereof the span is fifty thousand years* (Q. 70:4). Because of this fluid time-scale the Qur'anic account of the origin, as well as history, of the cosmos is based on a qualitative conception of time. Although this narration has certain outward resemblances with the Biblical account of creation, it is in essence very different from the account in Genesis, and this may be one reason why there has been no counterpart to the "young Earth" tradition in Islam.

Although the Qur'an does not explain how or when the cosmos was created, it does invite its readers to study the physical world. In fact, this Qur'anic invitation to observe the working of the cosmos is repeated with such insistence that it has now become commonplace to use these verses of the Qur'an to support the claim that the cultivation of modern science is a religious obligation for Muslims—a duty prescribed by none other than the Qur'an itself. Whether true or not, this simplistic approach does not do justice to the purpose of the Qur'anic invitation, for the Qur'an invites its readers to observe the order and regularities of the universe for the express purpose of understanding the realities that lie beyond the physical realm. This invitation to observe the physical cosmos is often accompanied by an emphatic reminder that observable order and regularities are a sign of the presence of the one and only Creator. The order observable in the physical cosmos is a testimony to Divine omnipotence, power, and wisdom.

The Qur'anic description of the world played a central role in the emergence of the different cosmographies in Islamic thought. These cosmographies describing main features of the cosmos developed through a complex process involving numerous currents of thought, including Arabic translations of Greek philosophical works, the interplay between various schools of thought within the Islamic philosophical tradition, and theological debates concerning the nature of God, His attributes, His relationship with the world, and other similar issues that had emerged much before the translation movement out of the internal dynamics of the Muslim community. These issues were not merely intellectual questions arising from the interpretation of the Qur'an but also had political, theological, and social dimensions. Debates on these questions gave rise to various schools of thought that eventually solidified into two main schools: the Mu'tazilah and the Asha'irah, both of whom were interested in cosmology and formulated a comprehensive theory of creation. In general, it was recognized that the physical cosmos exists within a larger scheme of creation that includes various levels of existence, including the nonphysical, and cannot be separated from this context. The cosmographies of the Sufis, in particular, describe the physical world in terms of degrees of being and existence.

The cosmographies that emerged in Islamic thought after the translation movement were dominated by debates over the question of the eternity of the world or its *ex nihilo* creation in time. In many of today's mainstream works dealing with the question of creation and eternity of the world in Islamic thought, battle-lines are drawn between Hellenized Muslim philosophers and their adversaries, the so-called orthodox thinkers, and the entire debate is shown to have emerged out of a crisis produced by the translation movement. In reality, the matter is much more nuanced. For instance, many cosmographies divided the physical world into the celestial and terrestrial regions, just as Aristotle did, but this did not imply an all-encompassing acceptance of Aristotelian thought. Even the most Hellenized philosophers of Islam (Ibn Sina and Ibn Rushd) had to recast the Aristotelian cosmos and his concept of its eternity, though they accepted the eternity of the world.

This modification of the Aristotelian cosmos is not merely a clever way of restating the same thing. For instance, the actual substance of which the physical cosmos is made was conceived by Aristotle as "matter,, an abstraction that could only be reached by means of a thought experiment: in his *Metaphysics,* after stating that substance is "that which is not predicated of a subject, but of which all else is predicated," he says that this statement itself is obscure, and

> further, on this view, *matter* becomes substance. For if this is not substance, it is beyond us to say what else is. When all else is taken away, evidently nothing but matter remains. For of the other elements some are affections, products, and capacities of bodies, while length, breadth, and depth are quantities and not substances. For a quantity is not a substance; but the substance is rather that to which these belong primarily. But when length and breadth and depth are taken away, we see nothing left except that which is bounded by these, whatever it be; so that to those who consider the question thus matter alone must seem to be substance. By matter I mean that which in itself is neither a particular thing nor of a certain quantity nor assigned to any other of the categories by which being is determined. (Aristotle 1984, 1625)

This description came under attack as early as the second half of the eighth century. Jabir bin Hayyan, for instance, declared this conception of matter to be "nonsense," writing no doubt in the tradition of Plotinus, who had called it a "mere shadow upon shadow":

> [You believe that] it is not a body, nor is it predicated of anything that is predicated of a body. It is, you claim, the undifferentiated form of things and the element of created objects. The picture of this [entity], you say, exists only in the imagination, and it is impossible to visualize it as a defined entity . . . all of this is nonsense. (Haq 1994, 53)

Likewise, Aristotle's prime matter, which he thought to be eternal and indestructible, was not accepted in the Islamic tradition by the majority of philosopher-scientists. In fact, on closer scrutiny we find that the many similarities between Aristotelian cosmological tradition and Islamic cosmological schemes are superficial; underneath there are profound differences between the foundational ideas of the two traditions. As seen earlier, even those philosophers who accepted the eternity of the world a la Aristotle did not accept the Aristotelian system in its totality; rather, they devised entirely new conceptual schemes. Ibn Sina is a case in point, and we will examine his ideas along with their rebuttals by other scholars in the next chapter.

The justification for drawing absolute battle-lines between philosophers and theologians weakens when we find that though many Muslim philosophers believed in the eternity of the world there were also notable exceptions. Al-Kindi, who is universally recognized as the first true Muslim philosopher, rejected the eternity of matter and the universe, despite the deep influence of Aristotle and Plotinus on his thought. In his treatise *On First Philosophy* al-Kindi uses the word *ibda'* (which means to begin something out of nothing) to denote creation *ex nihilo*. Al-Kindi also develops three arguments for the creation of the universe: (i) an argument from space, time, and motion; (ii) an argument from composition; and (iii) an argument from time (Craig 1979, 56).

GEOGRAPHY, GEODESY, CARTOGRAPHY

On the clear night of May 24, 997, a twenty-three-year-old man was standing outside the Central Asian town of Kath, situated on the river Oxus, waiting for the eclipse of the moon to begin. Hundreds of miles away, another man by the name of Abu'l Wafa (d. 997 or 998) was waiting for the same lunar eclipse to begin in Baghdad. The two men had arranged to use the eclipse of the moon as a time signal to calculate the difference in longitude between Kath and Baghdad. The first man's name was Abu Rayhan al-Biruni. He was born on September 4, 973 CE, in Khwarizm (now in Uzbekistan) and he died in Ghazna (now Ghazni, Afghanistan) around 1050. His place of birth, which now bears his name, was in the environs of Kath, located on the eastern bank of the river Oxus (whose original name is Amu Darya), northeast of Khiva. Jurjaniyya (modern Kunya-Urgench, Turkmenistan), the other main city of the region northwest of Khiva, lay on the opposite side of the river. Al-Biruni had spent a good deal of time in Jurjaniyya during the early part of his life and had begun his scientific training at an early age. "He had studied with the eminent Khwariziam astronomer and mathematician Abu Nasr Mansur. At the age of seventeen he used a ring graduated in halves of a degree to observe the meridian solar altitude at Kath, thus inferring its terrestrial latitude" (Kennedy 1980, 147–58).

Like many men of learning of his time, al-Biruni was interested in a wide range of subjects, including astronomy, astrology, applied mathematics, pharmacology, and geography. "He was not ignorant of philosophy and the speculative disciplines, but his bent was strongly toward the study of observable phenomena, in nature and in man. Within the sciences themselves, he was attracted by those fields then susceptible of mathematical analysis ... about half his total output is in astronomy, astrology, and related subjects, the exact sciences par excellence of those days" (Kennedy 1980, 151–52). This man, honored by his contemporaries by the honorific title of "the Master" (*al-Ustadh*), but "unknown to the medieval West, except perhaps by the garbled name Maître Aliboron" (Kennedy 1980, 156), has left us nineteen works on geography, geodesy, and mapping theory referenced in various other works.

Al-Biruni's works on geography were written at a time when this science had already been well established in the Muslim world. Various works by pre-Islamic Egyptian, Indian, Greek, and Persian geographers had been translated into Arabic or were being translated. It was a time of great discovery and expansion. New and improved techniques were making sea vessels safer and capable of longer journeys. Muslim traders, scientists, scholars, and preachers were traveling the length and breadth of the ancient trade routes, new cities were coming into existence, and there was a great interest in recording and authenticating geographical coordinates of distant places. At least some of this interest in geography was spurred by religious requirements such as performing Hajj, the annual pilgrimage that required traveling to Makkah. Many of these geographers were also historians, astronomers, mathematicians, chroniclers, and scholars of religion. Abu Rayhan al-Biruni is an excellent example of such a scholar-scientist, as his only work extant today, *The Book of the Determination of the Coordinates of Positions for the Correction of Distances between Cities*, is a mine of information not only about geography but also cosmology, history, religious practices, social customs, the political and economic situation of the time, relationships between different scientists, debates between scientists and scholars of his time, and many other subjects (al-Biruni tr. 1967)

Al-Biruni did not write his book in isolation; like his other scientific works, his geographical treatise emerged out of a vibrant tradition that had already gone through numerous substantial changes since its emergence in Islamic civilization. At the dawn of Islam, Arabs had a practical knowledge of the geographical areas through which they traveled or from where pilgrims came to Makkah. They also had a general conception of the nature of the Earth, which had been part of their folklore for centuries. This

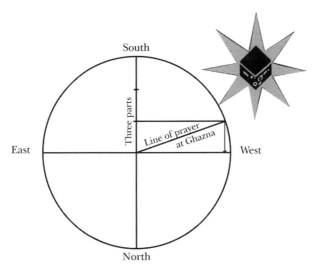

The need to determine the direction of Ka'bah (*qiblah*) played an important role in the development of many branches of Islamic scientific tradition, including geography and astronomy. Al-Biruni described a simple method for architects and artisans to determine *qiblah* from Ghazna in his *Kitab Tahdid al-Amakin*. The method involves drawing lines on a polished stable surface of a circle. © Al-Qalam Publishing.

knowledge however soon became inadequate for religious as well as practical reasons. The religious aspect of this inadequacy was primarily (but not entirely) due to a certain verse of the Qur'an revealed to the Prophet approximately sixteen months after his migration to Madinah. This verse (Q. 2:144), sometimes called the "verse of the changing of *qiblah*," commanded the Prophet and the believers to *turn thy face toward the inviolable house of worship* [the Ka'bah]; *and wherever you all may be, turn your faces toward it*. This religiously binding requirement to face the Ka'bah for the obligatory prayers five times a day was to give birth to a new dimension of geography. This "sacred geography," as it is sometimes called (King 1999, 51), is an entirely Islamic subbranch of geography that also spurred the development of a host of allied sciences such as mathematics, trigonometry, cartography, and mathematical geography. The need to determine the *qiblah* was easily met in Madinah, where everyone knew that the direction of Makkah was due south, but as soon as Muslim armies started to cross the frontiers of Arabia this became an urgent issue. One can imagine a Muslim army arriving in a remote town in Iran after crossing numerous mountains, hills, and deserts and having lost all sense of direction, save whatever could be gathered from the movement of the sun and the stars. Men in this army would have an urgent need: before the time for the next obligatory prayer they would need to know the direction of the Ka'bah. What could they do?

The initial solutions were approximate. During the seventh and the eighth centuries, when new mosques were being built in towns as far apart as Marv in Central Asia and the picturesque Seville in al-Andalus (Spain), Muslims had no truly scientific method for finding the correct direction toward the Ka'bah and so relied on folk astronomy. This told them that the rectangular base of the Ka'bah was astronomically aligned, with its major axis pointing to the rising of the star Canopus and its minor axis toward the extreme rising of the moon at midsummer and its setting at midwinter (King 1999, 49). From the ninth century onward, more scientific methods began to appear.

The need to determine the *qiblah* was, however, not the only reason for Muslims to develop a keen interest in geography and allied sciences. In the flow of its narrative, the Qur'an mentions several ancient nations and places that had incurred God's wrath because of their persistent opposition to prophets who came to guide them. As Muslims conquered new areas and came across old ruins, they attempted to identify them and see which of these had been mentioned in the Qur'an. These geographical studies became an integral part of the commentaries on the Qur'an. The Qur'an also mentions certain mountains (e.g., the mountain where Noah's boat came to rest; the mountain where Moses was called by God; the blessed mount of Tur), seven skies, and seven earths. These beame yet another direct reason for the emergence of those branches of geography that were concerned with the shape of the Earth, its extent, and its topography.

Thus, by the time of al-Biruni there had already developed several schools of thought that used different frameworks of inquiry for studying the Earth and its features. Translations brought fresh perspectives. During the reign of the Abbasid caliph al-Mansur (753–75), the Sanskrit treatise *Surya-Siddhanta* was translated into Arabic. Another work of considerable influence that was translated into Arabic at this time was the *Aryabhatiya* of Aryabhata of Kususmapura (b. 476), in which the author proposed that the daily rotation of the heavens was only an apparent phenomena caused by the rotation of the earth on its own axis, and, further, that the proportion of water and land on the surface of the Earth was equal (Ahmad 1991, 577). Another early influence came from Persia, where the notion of seven *kishwars* (*Haft Iqlim*) was predominant. In this scheme, the world was divided into seven equal geometric circles, each representing a *kishwar*. Persian maritime literature was also influential in the development of geographical studies in the Islamic tradition.

With the translation of Claudius Ptolemy's *Geography* into Arabic, Greek influence becomes apparent. Ptolemy's work was translated into Arabic

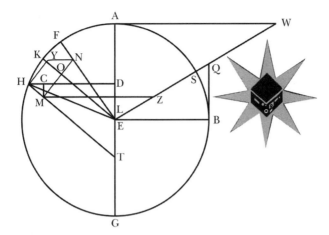

Various mathematical sciences developed to aid astronomy. In this diagram, al-Biruni shows how the cotangent of the displacement of the azimuth of the *qiblah* at Ghazna from the south point can be determined. The drawing is based on his *Kitab Tahdid al-Amakin*. © Al-Qalam Publishing.

several times during the Abbasid period, and each time it was translated it appeared with a new critique of the data. As Muslim domains expanded, new data too was added. In addition to Ptolemy, the *Geography* of Marinos of Tyre (*ca.* 70–130), the *Timaeus* of Plato, and the *Meteorology, De caelo,* and *Metaphysics* of Aristotle influenced the development of geography in the Muslim world. The translated material produced a great deal of activity, and by the beginning of the ninth century the science of geography was firmly established. It received further impetus from Caliph al-Ma'mun (813–33), who had a personal interest in geography. During his reign, "the measurement of an arc of a meridian was carried out (the mean result gave 562/3 Arab miles as the length of a degree of longitude, a remarkably accurate value); the astronomical tables called *al-Zidj al-mumtahan* (*The verified tables*) were prepared... [and] a World Map, called *al-Surat al-Ma'muniyya* was prepared" (Ahmad 1991, 578).

Many Muslim geographers were also philosophers, mathematicians, astronomers, mathematicians, and, more important for our discussion, scholars of religion. Among al-Kindi's 270 known works, for instance, there are several treatises on geography, astronomy, logic, metaphysics, and psychology. Other scientists who wrote on geography during the eighth and the ninth centuries include al-Fazari (eighth century), al-Khwarizmi (d. *ca.* 847), al-Farghani (d. after 861), al-Balkhi (d. 886), and, most important, Ibn Khurradadhbih (d. 911), who may have initiated the tradition of works generally entitled as *al-Masalik wa'l mamalik* (*Highways and Countries*) with a work by that title written in 846 and revised in 885. Ibn Khurradadhbih, a man of great learning, was the director of the Post and Intelligence Department in Baghdad, and his job provided him with opportunities to travel and encounter a great deal of geographical material.

Owing to the work of these eighth- and ninth-century geographers, two distinct genres appeared: the first dealt with the Muslim world of the time; the second described the geographical features of the entire known world. Works of the first kind included information on topography and road-systems of the Muslim world, while the second produced maps and general descriptions of the entire world. It is of interest for our discussion that not all geographical works used the same system to describe physical, human, and economic geography. Some presented their material using the cardinal directions as their reference points while others used the Persian system of *Iqlims* (regions). The latter took Makkah as the center of the world. The works of al-Istakhari (*fl*. first half of tenth century), Ibn Hawqal (*fl*. second half of tenth century), and al-Muqaddasi (d. 1000) belong to this category, and they are sometimes said to belong to the Balkhi School (named after Ibn Sahl al-Balkhi, d. 934). This School "gave a positive Islamic colouring to Arab geography" (Ahmad 1991, 581), and introduced innovations such as the element of perspective in cartography. They also gave Makkah the central position in their geographical representations.

Information used by Muslim geographers in their books and maps increasingly relied on first-hand accounts. Many works corrected previous Greek, Persian, Indian, and earlier data of Muslim geographers. Al-Mas'udi, for instance, "questioned [the] Ptolemaic theory of the existence of a terra incognita in the southern hemisphere, according to which the Indian Ocean was believed to be surrounded by land on all sides except in the east, where it was joined with the Pacific by a sea passage. He says he was told by the sailors of the Indian Ocean (*al-bahr al-habashi*) that this sea had no limits toward the south" (Ahmad 1981, 172). Travel accounts often provided an excellent source of information for geographers. As maritime travel increased, many new books were written to describe oceans and seafaring.

Encyclopedic works started to appear with the accumulation of data. These included world geographies, geographical dictionaries, maritime literature giving details of oceans and coastal regions, compilations specific to various regions, and general travel accounts. The most famous of the last category are *The Travels of Ibn Jubayr* (d. 1217) and Ibn Battuta (d. 1377). Abu'l Fida (d. 1331) has left us an outstanding work called *Taqwim al-buldan*, a general geography with a prologue full of interesting observations such as the gain or loss of a day as one travels around the world and descriptions of various rivers, lakes, seas, and mountains.

One of the greatest geographers of the twelfth and thirteenth centuries was Yaqut al-Hamawi (d. 1229). Of Greek parentage, he was taken prisoner as a young boy and brought to Baghdad, where he accepted Islam, learned Arabic, and spent the rest of his life in traveling throughout the Muslim world. A man of outstanding learning, Yaqut left us numerous encyclopedic works, four of which have been discovered. Among them is his *Mu'ajam al-buldan*, which has achieved the status of a classic and is still used as a reference work by scholars in the Muslim world as well as in the West. Arranged in alphabetical order, the *Mu'ajam* has preserved a wealth of information not only on geographical positions, boundaries, and coordinates but also on scholars, artists, and scientists. Passionately given to detail, Yaqut's geographical work is intimately linked with history; he was equally concerned with correct orthography, because he was aware that slight sloppiness could lead to big blunders. His inspiration to compile such a geographical dictionary came from the Qur'an, as he himself writes in his introduction (al-Hamawi 1959).

The Islamic West (*al-Maghrib*) produced its own specific geographies based on original observations, translations, and travel accounts. Al-Idrisi (d. 1165) is famous for his *Kitab nuzhat al-mushtaq fi'l khtiraq al-afaq*, written at the request of Roger II, the Norman King of Sicily. The book is the key to a large silver planisphere that al-Idrisi had presented to the monarch and completed in 1154. The book was illustrated with maps of various regions, and the six surviving complete manuscripts also contain the planisphere described in its introductory chapter.

The Ottomans translated many Arabic texts into Turkish and produced new works. They further rearranged old material, corrected geographical information, and added new observations. For example, Abu'l Fida's *Taqwim al-Buldan* was translated into Turkish by Sipahizade Mehemmed bin Ali (d. 1588), who supplemented and rearranged the material in alphabetical order. Turkish geographers also produced considerable new literature on marine geography and navigation. Seyyidi Ali Re'is bin Huseyn (d. 1562), also known as Katib-e Rumi, wrote a book on the Indian Ocean entitled *al-Muhit*, using the experiences of South Arabian sailors—some of whom had served as guides to Vasco de Gama on his voyage to Calicut (Taeschner 1991, 588). Piri Muhyi'l Din Re'is (d. 1554) produced a world map in 1513, for which he used as sources maps containing Portuguese discoveries up to 1508, and another map containing discoveries of Christopher Columbus during his third voyage (1498). He had obtained this map from a Spanish sailor who had voyaged with Columbus to America three times and who had been made a Turkish prisoner in 1501 at Valencia by none other than Piri Re'is's uncle, the famous naval hero Kemal Re'is (Taeschner 1991, 588). One of the most comprehensive geographical works of the early seventeenth century, written by Mustafa bin Abdallah, popularly known as Katib Khelebi or Haji Khalifa (1609–1657), also uses Muslim as well as at least one European source, the *Atlas Minor* (1621) of Gerhard Mercator (Taeschner 1991, 589).

Cartography, the science of production of maps and construction of projections and designs, became a basic need of the expanding Muslim world within the first generation. The administrative needs of the newly conquered lands required detailed descriptions, and early maps emerged on the basis of first-hand information of the new regions. This tradition was to receive a most direct impetus from the religious requirement already mentioned—the need to determine the *qiblah*.

Muslims may have received some Greek, Indian, and Pahlavi maps when the astronomical and geographical texts from these languages were translated into Arabic. We do not know when the first world map was constructed by Muslims, but we do know that the tradition of making these maps already existed in the ninth century, when a world map was constructed for Caliph al-Ma'mun (813–833) and named after him as *al-Surat al-Ma'muniyya*; al-Masu'di (d. 956) saw it and has left us the following account: "it depicted the universe with spheres, the stars, land and the seas, inhabited and barren (regions of the world), settlements of peoples, cities" (Ahmad 1997, 1078). Al-Khwarizmi's *Kitab Surat al-ard* (*The Book of the Shape of the Earth*) also contained coordinates of places (cities, rivers, mountains), and the original manuscript must have also contained maps, though they have not survived.

Cartography after the tenth century developed marked Islamic features, showing the influence of Islamic worldview in various realms such as politics and culture as well as various spiritual aspects of Islam. It emerged out of a new tradition in Islamic geography mentioned already—the so-called Balkhi Tradition, named after Abu Zayd Ahmad bin Sahl al-Balkhi (d. 934). Al-Balkhi's geographical work *Suwar al-Aqalim*, in which he described the geographical features of the Muslim world, dividing each province into an *iqlim*, was accompanied by maps that were copied and improved by al-Istakhari (d. 951). His work was then further improved upon by another excellent geographer, Ibn Hawqal (d. 977), who charted twenty-two maps, including a world map.

This new tradition of Islamic cartography differed from the Greco-Muslim tradition in many respects. Here the sacred city of Makkah occupies the central position; south is placed at the top while north is at the bottom, no doubt because of the reverence shown this sacred city (Ahmad 1997, 1079). In addition to the scientific aspects of the maps, consideration should be given to the characteristically Islamic aesthetics and color schemes, materials used for drawing maps (which ranged from brass to fine silk), and the abiding preoccupation with a directional grid oriented toward the *qiblah*, which perpetually reminded the believers of the Qur'anic concept of "the Straight Path." This physical geography was intimately connected with a nonphysical and "sacred geography in which directions, mountains, rivers, islands, etc. become symbols of the celestial world" (Nasr 1968, 99). In many cases, scientists have themselves told us that they were prompted to carry out their science because of the religious needs of the community, or because they felt duty-bound to correct certain wrong practices. The following passage from the celebrated *Mu'ajam al-Buldan* is but one such example:

> This is a book on the names of countries; on mountains, valleys, and plains; on villages, post-houses, and dwellings; on seas, rivers, and lakes; on idols, graven images, and objects of heathen worship. I have not undertaken to write this book, nor dedicated myself to composing it, in a spirit of frolic or diversion. Nor have I been impelled to do so by dread or desire; nor moved by longing for my native land; nor prompted by yearning for one who is loving and compassionate. Rather, I considered it my duty to address myself to this task, and, being capable of performing it, I regarded responding to its challenge as an inescapable obligation.
>
> I was made aware of it by the great and glorious Book, and was guided to it by the Great Tidings, wherein Allah said, glory and majesty to Him, when He wanted to manifest to His creatures His signs and warnings and establish their guilt by visiting upon them His painful wrath: *Have they not journeyed through the land? And have they not hearts to understand with, or ears to hear with? Surely as to these things their eyes are not blind, but the hearts which are within their breasts are blind.*
>
> This is a reproof to him who has journeyed through the world and has not heeded the warning, and to him who has contemplated the departed centuries and has not been deterred. (al-Hamawi 1959, 1–2)

MATHEMATICS

Around the year 825, a man in his twenties was sitting in a room in Baghdad, seeking help from God in writing a book that the Caliph al-Ma'mun (r. 813–833) had encouraged him to write—a concise book on "restoration" and "balancing" (*al-jabr wa'l-muqabalah*), which would be "useful in the calculation of what men constantly need to calculate [for their] inheritance and legacies, [their] portions and judgments, in their trade and in all their dealings with one another [in matters involving] measurement of land, the digging of canals, and geometrical [calculations], and other matters involving their crafts" (Khwarizmi 1989, 4). Three centuries later, this book was partially translated into Latin by Robert of Chester (*fl. ca.* 1150) as *Liber algebras et almucabola*; shortly afterwards, Gerard of Cremona (1114–1187) retranslated it as *De jebra et almucabola*, "introducing into Europe a science completely unknown till then, and with it, a terminology which was still capable of growth but already completely developed. This discipline was called by the two technical terms which appear in the titles of the first Latin translations, until the time when Canacci (fourteenth century) began to use only the first one, *algebra*" (Vernet 1997, 1070). In this manner, the title of the young man's book inaugurated a new branch of science, algebra. This was, however, not the only etymological contribution of this man; the Latinized version of his own name, al-Khwarizmi, would introduce into Latin and later into English the word algorithm, which is in common use today in computing science and mathematics. This singular distinction is perhaps consistent with al-Khwarizmi's hope, for in the introduction of his book *On Restoration and Balancing* he had written,

The learned of the times past and of nations which have ceased to exist were constantly busy writing books on various branches of science and knowledge, thinking of those who would come after them, hoping for a reward commensurate with their abilities, trusting that their endeavor would be acknowledged ... and relying on the purity of my intention and hoping that the learned will reward it by asking for me in their supplications, the most excellence of the Divine mercy, in requital of which may the choicest blessings and the abundant bounties of God be theirs. (Khwarizmi 1989, 4)

Al-Khwarizmi (b. *ca.* 780, d. after 847) had most probably come to Baghdad from Khwarizm. His contributions to arithmetic, algebra, geography, and astronomy were to play an important role in the subsequent development of these sciences, both in the Muslim world as well as in Europe. His book on arithmetic, *The Book of Addition and Subtraction by the Method of Calculations of the Hindus*, introduced the use of the Hindu numerals 1–9, the number zero, and the place value system still in use today. Within a century of its writing, al-Khwarizmi's work was used by Ahmad al-Uqlidisi (d. 980) for his *Book of Chapters*, in which he invented decimal fractions. Later, these two works were used by Yahya al-Maghribi (d. 1283) to find the roots of numbers and by Ghiyath al-Din Jamshid al-Kashi (d. 1429) to express the ratio of the circumference of a circle to its radius as 6.28318530717955865, a result correct to sixteen decimal places (Berggren 1986, 7).

Numbers we now use are made up of nine digits and zero—or *sifr*, an Arabic word from which the English word *cipher* is derived through the French and Spanish; the word *zero* is also derived from the Arabic *sifr* via Italian, which received it through the middle Latin word *zephirum*. These numbers have come to us from the Hindus, but they did not use this system to represent parts of the unit by decimal fractions, as that was an invention of Muslim mathematicians. We do not have al-Khwarizmi's book on arithmetic, but another early work, *Usul Hisab al-Hind* (*Principles*

Statue of al-Khwarizmi (*ca.* 847) in Samarqand, Uzbekistan. Al-Khwarizmi's book *al-Jabr wa'l muqabla* inaugurated the science of algebra. He made many original contributions in geography, mathematics, and astronomy. © Al-Qalam Publishing.

of Hindu Reckoning), written some 150 years after al-Khwarizmi's treatise, gives us an insight into the history of the development of decimal arithmetic. This work of an accomplished astronomer, Kushyar bin Labban (*fl. ca.* 1000), is in two sections and is supplemented by a chapter on the cube root. Kushyar uses a Babylonian system for fractions, a dust board for calculations, and *dirhams* as units of currency.

Al-Khwarizmi was not alone in making original contributions to the mathematical sciences; in the same century in which he lived, a number of other scientists produced works on different branches of mathematics. These include the famous philosopher al-Kindi (d. 873), his student Ahmad al-Sarakshi (*fl.* ninth century), al-Mahani (*fl.* ninth century), who was especially known for his study of Archimedes' problem, and the three sons of Shakir ibn Musa—Muhammad, Ahmad, and Hasan. During the next century, during which some of the most important and refined translations were made, Thabit ibn Qurrah translated the *Conics* of

Apollonius, many treatises of Archimedes, and the *Introduction to Arithmetic* of Nicomachus.

Like numbers, the decimal fractions we now use to represent fractions is an original contribution of the Islamic scientific tradition. This is clear from *The Book of Chapters on Hindu Arithmetic*, written in Damascus in the years 952–953 by Abu'l Hasan al-Uqlidisi. Al-Uqlidisi introduced decimal fractions in the second part of his work in the section on doubling and halving numbers; while his use of decimal fractions was somewhat ad hoc, two centuries later Ibn Yahya al-Maghribi al-Samawal (*fl.* twelfth century) introduced them within a theory of numbers, though still without naming the device. By the early fifteenth century, decimal fractions had received both a name and a systematic exposition, as we see in the work of Jamshid al-Kashi, an extraordinary mathematician and astronomer who later joined the team of astronomers and mathematicians that had gathered in Samarqand at the invitation of Ulugh Beg (d. 1449), the grandson of Timur (d. 1405) (Berggren 1986, 36).

No account of mathematics in Islamic civilization can be complete without the mention of Omar Khayyam (1048–1131), most known in the West for his *Rubaiyat* (*Quatrains*). Khayyam wrote an undiscovered treatise called *Problems of Arithmetic* (*Mushkalat al-hisab*), a key treatise on cubic equations (the *Risala*), a lengthy commentary on Euclid, and many other works on astronomy, music, arithmetic, and algebra, in addition to his better-known poetic and philosophical works. Sometime after 1070 he became the head of the team of the most distinguished astronomers of the eleventh century, who compiled *Zij Malik-Shahi* (*Malik-Shah Astronomical Tables*) at the observatory in Isfahan, a city where he spent the eighteen most peaceful years of his life. The small portion of this work that has survived consists of tables of ecliptic coordinates and of the magnitudes of the 100 brightest fixed stars. Around 1079 he proposed a reform for the calendar then in use. According to his reform, "the average length of the year was to be 365.2424 days (a deviation of 0.0002 day from the true solar calendar), a difference of one day thus accumulating over a span of 5,000 years. (In the Gregorian calendar, the average year is 365.2425 days long, and the one-day difference is accumulated over 3,333 years.)" (Youschkevitch and Rosenfeld 1980, 324).

That Islam was instrumental in at least some of these developments can be seen from the fact that al-Khwarizmi devoted the second half of his *algebra* to examples for calculating inheritance (for which the theoretical ratios are supplied by religious law) and *zakah* (the obligatory charity every Muslim gives from his wealth if it exceeds a certain amount). Both the division of inheritance and the calculation of *zakah* can be a complicated affair, and the development of corresponding mathematical formulae requires a full understanding of the religious laws involved. In addition, a great

Mausoleum of Omar Khayyam (1048–1131) in Neshapur, Iran. In addition to being a famous poet, Khayyam was an accomplished mathematician and astronomer. In 1079, he proposed a reform for the calendar in use. The average length of the year in his new calendar deviated by 0.0002 day from the true solar year. © Al-Qalam Publishing.

deal of applied mathematics was required to solve astronomical problems associated with other religious practices, as we shall discuss later in this chapter. Similarly, mathematics was an indispensable tool in sacred geography, as we have already seen.

These are, however, only the most apparent dimensions of the relationship between Islam and mathematics. There is a much deeper and foundational aspect to this relationship involving metaphysical realities expressed through numbers. Each number from zero to nine, in addition to its numeric character, also represents a geometrical figure and, hence, a "personality." Each letter of the Arabic language was also assigned a number, and a complete science (*jafr*) based on the numerical values of the letters emerged. This has numerous dimensions, ranging from mystical interpretations of the Qur'an to the tradition of writing verses from which the date of death of the deceased can be calculated. Thus, numbers were not merely symbols of quantities; through geometrical shapes on the one hand and the science of *jafr* on the other, they represented numerous spiritual and aesthetic aspects of the created order.

ASTRONOMY

Two verses of the Qur'an played a key role in establishing a nexus between astronomy and Islam. The first established the lunar year as consisting of twelve months, four of which were specified as sacred (Q. 9:32); the second (Q. 2:149–50) changed the direction of the *qiblah* from Jerusalem toward the Ka'bah in Makkah, requiring Muslims to face this direction for the ritual prayers and certain other acts of worship. The Qur'anic injunction to establish *salah*, the ritual prayers, at specific times also caused the development of a special branch of religious astronomy called *ilm al-miqat*, the science dealing with three distinct aspects requiring astronomical solutions: the direction of *qiblah*, the determination of the times for prayers, and the visibility of the new moon. We have a precise definition of this science by a fourteenth-century Egyptian scholar, Ibn al-Akfani, who was the author of an encyclopedia and several works on medicine. He states,

The science of astronomical timekeeping is a branch of knowledge for finding the hours of the day and night and their lengths and the way in which they vary. Its use is in finding the times of prayer and in determining the direction in which one should pray, as well as in finding the ascendant and the right and oblique ascensions from the fixed stars and the lunar mansions. This science is also concerned with shadow lengths and the altitudes of celestial bodies, and with the orientation of one city from another. (King 2004, 648)

Initially approximate methods based on folk astronomy were used to determine the direction and times of prayers. These methods used astronomical phenomena visible to the naked eye, the direction of winds, the position of stars, and the like. But as astronomical research progressed, more sophisticated methods came into existence. By the middle of the ninth century, sacred astronomy had become fully established. Numerous scientists contributed to the development of this science. Among them, al-Khwarizmi (d. 847) and al-Battani (d. 929) hold special stature for proposing new tables based on the difference of longitudes between Makkah and a given place. Al-Battani's description of astronomy provides an insight into the high esteem in which this branch of science was held by Muslims. In the beginning of his *Zij al-Sabi*

he describes astronomy with such phrases as 'the most noble of the sciences in rank',' elevated in dignity', 'illuminating to the soul', 'pleasing to the heart'... 'a field of endeavor with an invigorating effect on the intellect' and 'as sharpening the faculty of reflection'... field which makes possible the knowledge of the length of the year, the months, and different times and seasons, the lengthening and shortening of day and night, the positions of the sun and the moon as well as their eclipses, and the courses of the planets in their direct and retrograde motions, the alternations of their forms, and the arrangement of their spheres; and he asserts that these lead people, who reflect deeply and persistently, to the proof of the unity of God and to the comprehension of His majesty, to His immense wisdom, infinite power, and to a grasp of the excellence of His act. (Sayili 1960, 15–16)

Further developments in this field were led by scientists such as Habash al-Hasib (d. 864), al-Nayrizi (d. 922), and Ibn al-Haytham (d. 1040). Al-Biruni (d. 1050) used spherical trigonometry to provide solutions. During the thirteenth century, new formulations appeared due to the work of astronomers such as Abu Ali al-Marrakushi (*fl.* 1281), whose method was probably used by the Damascene *muwaqqit* al-Khalili (*fl.* 1365) to compute his extremely developed and accurate *qiblah* table (Samsó 2001).

Research in astronomy as well as those disciplines of science that were required for astronomical research (mathematics, trigonometry, etc.) was directly related to Islam in that it was needed by the community, but in addition to this utilitarian purpose astronomical research was carried out for its own sake throughout the Muslim world. A related development was the appearance of brass maps of the world, with various localities placed on grids. This was an art that required the knowledge of sophisticated mathematics and geometry. The discovery of two such world maps for finding the direction and distance to Makkah has helped to push the date of the decline of science in Islamic civilization well beyond initial estimates.

These two maps are engraved on a circular plate and are believed to have been made in the middle of the second half of the seventeenth century (King 1999, 199).

Astronomical time-keeping also led to the development of *miqat* tables computed on the basis of the coordinates of a given locality. One of the earliest *miqat* tables is attributed to Ibn Yunus (d. 1009), whose work provided the basis of numerous subsequent tables that were used in Cairo until the nineteenth century (Samsó 2001, 212). By the middle of the twelfth century, most cities had official *miqat* tables and, in certain big cities, a special official had been appointed for this purpose. Ibn al-Shatir (d. 1375) is said to have held this office in Damascus. The third standard problem of *miqat*, the prediction of crescent visibility, which determined the beginning of a new Islamic month, received focused attention of Muslim astronomers throughout the period under consideration and remains an area of special interest even now. We have an inside narrative from one of the most celebrated Muslim astronomers that testifies to these abiding concerns of Islamic astronomy. This is in the form of a letter by Ghiyath al-Din Jamshid Mas'ud al-Kashi to his father, written a few weeks after his arrival in Samarqand to take part in the construction of a new observatory. This letter, fortunately preserved for posterity by his father, is also an intimate source of rich details on the nature of science in the Islamic civilization in the fifteenth century—a time once considered to be barren!

Al-Kashi begins his letter by thanking God for his many favors and blessings, then apologizes to his father for not writing earlier. He speaks of his preoccupation with the observatory and tells him how he had been received by Ulugh Beg, a ruler whom he describes as extremely well-educated in the Qur'an, Arabic grammar, logic, and the mathematical sciences. He relates an anecdote about Ulugh Beg, that one day, while on horseback, he computed a solar position correct to the minutes of an arc. He then tells his father that upon his arrival he was put to test by more than sixty mathematicians and astronomers who were already working in Samarqand at the Ulugh Beg complex. He was asked to propose a method for determining the projections of 1022 fixed stars on the rete of an astrolabe one cubit in diameter; to lay out the hour lines on an oblique wall for the shadow cast by a certain gnomon; to construct a hole in a wall to let in the sun's light at, and only at, the time of the evening prayer; and to find the radius, in degrees of an arc on the earth's surface, of the true horizon of a man whose height was three and a half cubits. All these and other problems, which had baffled the entourage, al-Kashi tells his father, were solved by him "without much difficulty, earning [him] respect and honor" (Kennedy 1960, 3–4).

In addition to mathematical astronomy and the *miqat* tradition, Islamic astronomy has also left us a rich legacy of observatories and astronomical

Statue of Ulugh Beg (d. 1449), the ruler of Samarqand who built one of the most important scientific centers of the fifteenth century. © Al-Qalam Publishing.

instruments. In fact, observatories, hospitals, madrassahs, and public libraries are four characteristic institutions of Islamic civilization. The first observatory may have been constructed by Muslims during the Umayyad period (661–750). We have definite information of a systematic program of astronomical observations from the time of al-Ma'mun, who was the patron of this research carried out at the Shammasiyya quarter in Baghdad (in 828–829) and at the monastery of Dayr Murran on Mount Qasiyun in Damascus (831–832) (Sayili 1960, 50–56).

The most advanced astronomical research in Islamic scientific tradition may have been carried out at Maragha in western Iran between the middle of the thirteenth and fourteenth centuries—a period that has been called the "Golden Age" of Islamic astronomy (Saliba 1994, 252). The work of four astronomers—Mu'ayyad al-Din al-Urdi (d. 1266), Nasir al-Din al-Tusi (d. 1274), Qutb al-Din al-Shirazi (d. 1311), and Ibn al-Shatir (d. 1375)—is particular important. They belong to what has been called the "Maragha school" (Roberts 1966), and their work was in continuation of a tradition of criticism of Ptolemy that had begun as early as the eleventh century. The work of the Maragha School was revolutionary in the history of astronomy, and paved the way for a complete overhaul of Ptolemy's model. Ptolemy had described the movements of the planets, including the Sun and the Moon, on epicyclic spheres that were in turn carried within the thickness of other spheres that he called deferents. He represented these spheres by circles. Ibn al-Haytham (d. 1048) and Abu Ubayd al-Juzjani (d. *ca.* 1070) noticed several contradictions in Ptolemy's model of the universe. Ibn al-Haytham noted in his landmark work, *al-Shukuk ala Batlamyus* (*Doubts Concerning Ptolemy*), that one cannot assume there is a sphere within a physical universe that would move uniformly around an axis without passing through its center (Saliba 1994, 251). He pointed out that the Ptolemaic equant was in direct violation of this principle. Ibn al-Haytham concluded that Ptolemy's description could not be a true description of the physical universe and hence should be abandoned for a better model.

This tradition of critical examination of Ptolemaic model continued in the western part of the Muslim world with important contributions made by Andalusian astronomers such as al-Bitruji (*ca.* 1200), Ibn Rushd (d. 1198), and Jabir bin Aflah (*ca.* 1200). However, it was in Maragha that a revolutionary change took place: in 1957 Victor Roberts showed that the lunar model of Ibn al-Shatir (d. 1375) was essentially identical to that of Copernicus (1473–1543). Since then many other historians of science have conclusively shown that Copernicus essentially used the work of Muslim astronomers, although the route of this transmission still remains unclear (Kennedy et al. 1983). "The question therefore is not whether, but when, where, and in what form he [Copernicus] learned of Maragha theory" (Saliba 1994, 255). The work of these historians of science on the Maragha school has revolutionized our understanding of the nature of Islamic scientific tradition.

In addition to its useful religious functions, astronomy also served astrologers, who were generally condemned for their claim to have the

The madrassah of Ulugh Beg (d. 1449) in Samarqand, where astronomers and mathematicians of the fifteenth century made significant discoveries. © Al-Qalam Publishing.

knowledge of future events. This claim was in contradiction to the Qur'anic teaching that only God has knowledge of the future (Q. 27:64). The claims by astrologers, therefore, amounted to claiming a share in God's knowledge. In addition to the Qur'an, many sayings of the Prophet also condemned regarding the stars and their movements as sources of one's fortunes or misfortunes, and these led Muslim scholars to develop extensive criticism of astrology. Despite all this, astrology remained popular with rulers and the elite, and this sometimes produced tensions that spilled over into the related area of astronomy. This may have been the cause of the closing of the Istanbul observatory, but there were also other political motives behind that incident.

Astronomical research required the use of certain instruments. Muslims had inherited some knowledge of instrument making from Ptolemy's *Almagest*, but they invented many new instruments over the course of eight centuries, including observational instruments as well as analog computers. Quadrants, altitude sextants, semicircular, ruler instruments, and other observational tools were used to determine altitudes and azimuths; armillary instruments were used for measuring right ascensions and declinations and for longitudes and latitudes with respect to the ecliptic. Sextants and bipartite arcs, as a third type of observational instrument, were

The site of the Maragha Observatory in Iran. Advanced astronomical research was carried out during the thirteenth and the fourteenth centuries by astronomers such as Mu'ayyad al-Din al-Urdi (d. 1266), Qutb al-Din al-Shirazi (d. 1311), and Ibn al-Shatir (d. 1375). © Al-Qalam Publishing.

used to determine the angular distances between celestial bodies. Various accounts of these instruments provide insights into their improvements. Al-Biruni's previously cited work, *The Determination of the Coordinates*, also provides information about developments in various instruments. The use of the mural quadrant, for instance, was an important development in practical astronomy, and its accuracy was not surpassed until the use of optical instruments. The invention of the mural quadrant is generally attributed to Tycho Brahe and is named after him; recent scholarship has shown that the so-called Tycho's Mural Quadrant (or Tichonicus) was already in use in the Muslim world during the time of Nasir al-Din al-Tusi; Taqi al-Din's observatory in Istanbul had a mural quadrant of a 6 m radius, whereas the radius of Tycho's quadrant was only 194 cm (Dizer 2001, 248)

Other instruments developed or improved upon by Muslims include the armillary sphere, first described by Ptolemy but apparently never constructed until its use by Muslims. A variation on this instrument constructed at the Maragha observatory had five rings and an alidade instead of six rings. This increased the convenience of use without reducing accuracy.

The most important analog computer used by Muslims was the astrolabe. Its origins are definitely pre-Islamic, but it received sustained and focused attention by Muslims, who perfected its use and made many improvements in its design. "The ability of Islamic civilization to perfect what it inherited," observed Oliver Hoare, "and to endow what it made with beauty, is nowhere better expressed than in the astrolabe" (King 1999, 17). A concise and useful description of this widely used instrument may be helpful.

> An astrolabe is a two-dimensional representation of the three-dimensional celestial sphere. The rete, bearing pointers for various bright stars and a circle representing the ecliptic—the 'celestial' part of the instrument—can rotate over any of a series of plates for specific latitudes—these being the 'terrestrial' part of the instrument—marked with the horizon and meridian as well as altitude and azimuth curves. A sighting device on the back of the instrument enables one to measure the altitude of the sun or any star; one then places the appropriate mark on the rete on top of the appropriate altitude circle on the plate for the latitude in question. The instrument then shows the instantaneous configuration of the heavens with respect to the local horizon. (King 1999, 18–19)

Astrolabes were in use in the Muslim world as early as the time of al-Fazari, who died in 777. By the end of the eighth century, the making of astrolabes had become an important art in the Muslim world. Among the famous early authors who wrote treatises on the astrolabe are al-Marwarrudhi and his student Ali bin Isa, nicknamed al-Asturlabi. Al-Khwarizmi has also left us a compendium of numerous problems to be worked out with the astrolabe and a treatise on its construction (Dizer 2001, 257). The subsequent history of the astrolabe is a fascinating tale of the coordination and merger of various Islamic arts and handcraft with the practical needs of astronomy. Numerous sophisticated astrolabes made of wood, brass, and other metals exist in various collections worldwide. Many await proper study (King 1999, 17).

PHYSICS

In general, physics remained entrenched in the Aristotelian framework throughout the eight hundred years of the Islamic scientific tradition. It was thus conceived as a branch of science dealing with change. Change was studied in the general Aristotelian framework of form and matter, potentiality and actuality, and the four causes. This was the position of the Muslim Peripatetics, who followed Aristotle in their understanding of physics. They held a dominant (though not exclusive) position in the study of change.

The sextant of Ulugh Beg's observatory. The marble arc of about 600 m and a radius of about 40 m was used to make observations. Astronomical research carried out at Ulugh Beg's observatory produced a star catalog consisting of 1,018 stars. © Al-Qalam Publishing.

The astrolabe was the most important astronomical instrument used by Muslim astronomers and mathematicians. Of pre-Islamic origins, the astrolabe was greatly improved by Muslims. © Al-Qalam Publishing.

The Peripatetics were however opposed by scientists, philosophers, and religious scholars, who challenged Aristotle's views, though for different reasons. While independent scientists such as Abu Bakr Zakaria al-Razi and al-Biruni opposed their coreligionist Peripatetics on scientific and philosophical grounds, theologians such as Abu'l Barakat al-Baghdadi (d. 1023) and Fakhr al-Din al-Razi (d. 1209) opposed Aristotelian views from a religious and philosophical perspective and formulated a view of time, space, and causality distinct from Aristotle's. In this non-Aristotelian conception of nature one finds distinct Islamic characteristics, both in the way in which the creation of things was perceived as well as in the way change takes place. For instance, the theory of balance proposed by Abu'l-Fath Abd al-Rahman al-Khazini (*fl.* 1115–1130) in his major treatise *Kitab Mizan al-Hikma* (*The Book of the Balance of Wisdom*) deals with the concept of center of gravity in non-Aristotelian ways. He also continues on the work on hydrostatics and mechanics by al-Biruni, al-Razi, and Omar Khayyam.

This anti-Aristotelian current in Islamic philosophy and science produced new ideas about time, space, and the nature of matter and light, but tension between those who held to the Aristotelian view of nature and those who were anti-Aristotelian cannot be regarded as a tension between "Islam" and "science," as is sometimes stressed to emphasize the conflict model; rather, it is a tension mostly confined to the realm of philosophy, as it is mainly focused on those philosophical beliefs of Aristotle that opposed Islamic beliefs and had no scientific basis.

These opposing currents appeared in Islamic tradition at the beginning of the translation movement, in the middle of the eighth century, and became particularly strong in the middle of the eleventh century when almost the entire corpus of Greek philosophical and scientific works had been translated into Arabic.

Other branches of physics—such as optics, mechanics, and dynamics—were included in the standard works of Muslim scientists and philosophers, and though much of it remained Aristotelian in outlook and basic doctrine, non-Aristotelian concepts of matter, space, time, and causality were also present.

For example, the *mutakallimun* developed atomistic theories where even time was atomic and where the only true causality worked downward from God. And at the other extreme of religious respectability, al-Razi had an idea of absolute space that pushed towards the Newtonian view and was opposite Aristotle's 'place'-determining plenum. Certain epistemological and methodological issues were important concerns of dynamics: the possibility and legitimacy of abstraction (to empty space, to forceless conditions), the possibility or reliability of mathematical treatment. (Hall 2001, 319–20)

GEOLOGY AND MINERALOGY

Geology (as a science that deals with the dynamics and physical history of the Earth, the rocks of which it is composed, and the physical, chemical, and biological changes that the Earth has undergone or is undergoing) was of special interest to Muslim scientists and philosophers of the medieval times for two reasons: (i) it was a branch of science to which attention was diverted by the Qur'an itself and (ii) it was practically useful. What was observable on the Earth was a vast system of change over long periods of time, and this system was taken as a Sign of God. For Muslim scientists of the period between the eighth and the sixteenth centuries, the Earth was created by God for a fixed duration and for a definite purpose. It was a place of wonders and of observations that led to an understanding of Divine Wisdom, Power, and Mercy, with sustenance provided for every living creature on Earth. For these scientists, the Qur'anic descriptions of various processes such as the regeneration of water and of the Earth's *coming back to life after having been barren* (Q. 36:23) were observable realities that directed human reflection to the Creator. They studied various geological processes (such as weathering, erosion, and transportation), the stratigraphic arrangement of strata, and long geological spans of time within the context of the Qur'anic descriptions of creation.

The earth was considered to have a special position in the universe. Special attention was paid to the formation of rocks, mountains, the course of rivers, natural processes, minerals and lapidaries. Various stones and gems were studied for their medicinal properties, and many stones and minerals were converted into digestible form and were thus incorporated into the Islamic pharmacological tradition.

Certain works by Muslim scientists of the period under consideration offer remarkable examples of their observational abilities and formulation of geological theories. For example, al-Biruni has the following statement in his *Tahdid*:

> We do not know of the conditions of creation, except what is observed in its colossal and minute monuments which were formed over long periods of time, for example, the high mountains which are composed of soft fragments of rocks, of different colors, combined with clay and sand which have solidified over their surfaces. A thoughtful study of this matter will reveal that the fragments and pebbles are stones which were torn from the mountains by internal splitting and by external collision. The stones then wear off by the continuous friction of enormous quantities of water that run over them, and by the wind that blows over them. This wearing off takes place, first, at the corners and edges, until they are rubbed off and the stones finally take an approximate spherical shape. As a contradistinction to the mountains, we have the minute particles of sand and earth. (al-Biruni tr. 1967, 16)

Al-Biruni has also left us a work on precious stones, *The Book Most Comprehensive in Knowledge on Precious Stones*, which combines philosophical insights, geological information, history, pharmacology, comments on various rulers, the habits of some nations, and numerous other reflections in a compelling narrative. Al-Biruni describes various precious stones, narrates events from his travels, quotes poets, story-tellers, and philosophers, all the while describing the value or properties of precious stones and gems. He also gives scientific descriptions of the way certain precious stones are formed.

In addition to al-Biruni, many other Muslim scientists, such as al-Kindi, Ibn Sina, al-Mas'udi, and al-Idrisi have also left valuable works on geology and mineralogy. All of these works reflect a worldview anchored in the Qur'an and a belief system that takes the Earth as a Sign of the Creator.

OTHER BRANCHES OF SCIENCE

In addition to the already mentioned branches of science, there are others—such as zoology, veterinary science, alchemy, and various medical sciences—that had numerous direct and indirect relations with Islam. For instance, the Qur'an mentions a number of animals and birds by name and speaks of their benefits to humanity. Chapters 2 and 16 of the Qur'an are named after the cow and the honeybee, respectively. Horses are specifically mentioned in many verses. The first five verses of the hundredth chapter, for instance, describe charging horses in magnificent rhythm. Eating meat of certain animals is considered unlawful in Islamic Law. Many plants and fruits are also mentioned in the Qur'an. These verses of the Qur'an provided a certain theoretical framework for the cultivation of zoology, botany, and other related sciences

The words and practices of the Prophet also formed the basis of practical psychology which was greatly advanced by Muslim scientists and scholars. Practical psychology was also shaped by the Qur'anic verses dealing with the relationship between human condition and the Creator. *Indeed, the hearts receive tranquility by the remembrance of Allah*, we are told in a verse (Q. 13:28). The Prophet recommended specific supplications to the sick. It was his practice to visit the old and the sick and give them hope and joy. The verses of the Qur'an and the practice of the Prophet were used by Muslim scholars to develop a comprehensive framework for the practice of sciences related to health, preventive medicine, and psychological well-being.

Within the overall framework of Islamic medical sciences, the tradition of the "Prophetic Medicine" was obviously directly inspired by the teachings and practices of the Prophet of Islam. Studies in this particular branch of medicine were accompanied by an effort to preserve the exact words of the Prophet. His sayings dealing with health, sickness, hygiene, and other issues related to medicine contain specific references to diseases such as leprosy, pleurisy, and ophthalmia. He recommended remedies such as cupping, cautery, and the use of honey and other natural substances. This body of *hadith* on medical issues was systemized by religious scholars who were often practicing physicians. Thus literature on *Tibb al-Nabawi* (the "Prophetic Medicine") is a distinct genre in Islamic medical sciences and there exist numerous works dealing with various aspects of this tradition. One of the most celebrated work of this type is the *Tibb al-Nabawi* of Ibn Qayyim al-Jawziyya (d. 1350). In his book, Ibn Jawziyya provides general principles of health and sickness, reflects on the relationship between medicine and religion, enumerates the sayings of the Prophet concerning medicine and discusses the role of Divine Revelation in medicine (al-Jawziyya tr. 1998). He also specifically mentions various remedies recommended by the Prophet. Another related aspect of this tradition of Prophetic Medicine is the body of literature dealing with pharmacological studies on various herbs and other natural substances used or recommended by the Prophet. A whole branch of scientific research has been inspired by the teachings of the Prophet on the usage of these substances and these studies continue to this day.

In general, it is safe to say that the enterprise of science in Islamic civilization had numerous direct and indirect connections with the Islamic worldview. In some branches of science these connections were more obvious and identifiable; in others, Islam's particular conception of nature played a more indirect role. The next chapter explores some of these connections.

Chapter 4

The Mosque, the Laboratory, and the Market (the eighth to the sixteenth centuries)

UNDERSTANDING THE ISLAM AND SCIENCE NEXUS

It has been said in another volume in this series that "one's approach to science and religion interactions will depend on how one defines the central purposes and the appropriate boundaries of science and/or religion. The philosopher Immanuel Kant (1724–1804) and the extremely influential Protestant theologian Rudolph Bultmann (1884–1976), for example, insisted there could be no authentic interactions between the two" (Olson 2004, 1). Since the times of Kant and Bultmann, numerous other influential philosophers and scientists have reiterated the same opinion and, as Olson has noted, Bultmann's position has remained characteristic within the Christian tradition in the West from the beginning of the Renaissance to the present. It is in this historical context that Galileo Galilei's remark (borrowed from Cardinal Cesare Baronius) that "the Bible tells us how to go to heaven but not how the heavens go" is extremely relevant (Olson 2004, 2). One cannot, however, find a similar historical incident in the relationship between Islam and the scientific tradition that existed in Islamic civilization between the eighth and the sixteenth centuries.

The relationship between science and Christianity has also been redefined because of substantial changes to the role of religion in Western civilization since the Renaissance. A similar shift in the place of religion has not occurred in the Muslim world. Even the very definition of religion, as understood in contemporary Western civilization, is substantially different from how it is understood in the Muslim world.

As we begin an in-depth exploration of the Islam and science nexus, it is important to begin with Islam's self-definition as a "religion" rather than to use definitions proposed by Western thinkers such as Webster, who defines religion as "a set of beliefs concerning the cause, nature, and purpose of the universe, especially when considered as the creation of a superhuman agency or agencies, usually involving devotional and ritual observances, and often containing a moral code governing the conduct of human affairs." The Arabic word used for religion in the Qur'an and in other Islamic texts is *dîn* (from the trilateral root D-Y-N), which is a comprehensive term with multiple layers of meaning, including "to obey," "to be subservient to God," "a way of life," "Divine Law," "a pattern," and "Recompense." *Dîn*, thus, is not merely a set of beliefs and associated practices, but a way of being, a path that a traveler takes with a definite destination in mind. Moreover, it is a path that *transforms* travelers as they order their lives according to Divine guidance. While there is of course a set of beliefs associated with Islam, these beliefs are not merely abstract ideas; they have been presented in concrete form through the "Way" (*Sunnah*) of the Prophet of Islam, which constitutes the real-life model for travelers on the *dîn* of Islam.

We should also keep in mind that Islam's self-definition is not limited to the specific "religion" initiated by Muhammad in the seventh century; rather, the Qur'an considers Islam to be that path and way (*dîn*) that corresponds to, and is in harmony with, the innate nature of all human beings, *fitrah*—the pattern on which they are created. And since this innate nature of human beings is an unchanging characteristic, their *dîn* too has remained unchanged since the dawn of humanity. The Qur'an states that all messengers of God have brought the same message to humanity. What has varied in different manifestations of this *dîn* has been merely outward and secondary aspects: specific forms of worship, specific things and practices declared lawful and unlawful, specific rituals. The message given to Muhammad, according to the Qur'an, completed the cycle of revelation, confirming all previous revelations. "Religion" thus understood is not merely an inert set of beliefs and associated practices; rather, it is an ever-present consciousness in the deepest recesses of a person. The aim of *dîn* is to reestablish "the bond—the ligament between man and God which man lost at the Fall. Every religion is thus something like a rope thrown down from Heaven for fallen man to cling to" (Lings 2004, 1).

Three essential elements of Islam can be summarized as follows: (i) belief in one God, the Creator, Originator, and Sustainer of all things; (ii) belief in the veracity of all messengers appointed by God to deliver His message; and (iii) belief in the eventual end of all things, ushering in another kind of life in the Hereafter. These three cardinal elements of Islam, as primordial *dîn*, are denoted by three technical terms: *Tawhid* (Oneness of God), *Risalah* (Prophethood), and *Ma'ad* (Return). These are presented by the Qur'an in various ways and are supported through numerous demonstrational proofs from three realms: the cosmos, human history, and the human soul (*nafs*). Understood in the Qur'anic terms, *religion* refers to an existential reality permeating every existing being—from mighty mountains to tiny ants crawling in the vast desert. *All that is in the heavens and the earth extols Allah, the Mighty, the Wise*, as the Qur'an declares in its own characteristic manner (Q. 57:1).

This specific understanding of "religion" demands that the questions related to the Islam and science nexus be formulated according to Islam's self-definition. To be sure, there will be some overlap between these questions and those normally used to explore the relationship between science and Christianity (or, for that matter, any religion), but this overlap is a secondary, not essential, characteristic of this discourse.

The religion and science discourse in Christianity has also found common ground in the two Books of God—scripture and nature—on the basis of autobiographical anecdotes of scientists who have incorporated knowledge, methods, concepts, and ideas from scripture into their scientific works. This discourse thus takes the personal beliefs of a Newton (1642–1727), Boyle (1627–1691), or Einstein (1879–1955) to be indicative of porous boundaries between science and religion (Olson 2004, 3). This is used to build the case for their interaction. This argument can obviously be applied with much more emphasis to Islam and science, because, compared to the post–seventeenth-century era, scientists who made the Islamic scientific tradition possible during the eighth and the sixteenth centuries were more frequently also the authors of religious texts of an advanced nature. This argument, however, has little relevance, because—as opposed to the post-Renaissance scientists noted earlier—the Muslim scientists of the pre-seventeenth century era did not see their religion and their science as two separate entities.

Another problematic argument stemming from the universalization of the specific interaction between Western Christianity and science relates to the notion that religious rituals and celebrations related to the calendar are a "reason" for science and religion to interact, as science and scientists help religious institutions and authorities to prepare for these rituals and celebrations by furnishing required astronomical information. The need to determine the spring equinox for Easter, for instance, is considered a religious need that is fulfilled by science, and hence religion is shown to have an intrinsic need to support scientific research for its own needs (Olson 2004, 3). This argument can be superficially applied to Islam (as is often the case) with much more force, because in the case of Islam, not only is the annual cycle of rituals and celebrations dependent on astronomically determined times but also daily practices, such as the five obligatory prayers and fasting. These obligations are called "religious obligations" in a framework in which some obligations are "religious" and others are "nonreligious." Because, however, Islam considers itself a complete way of life (*al-dîn*) encompassing the entire spectrum of human activities—making even the most ordinary act of removing a harmful thing from the road a "religious" act—these categories and the arguments associated with them become superfluous. What is perceived as the "need of religion" being "fulfilled by science" is thus a conception foreign to Islam, which naturally perceives the needs of humanity in terms of its obligations towards the Creator, and so defines a way of life for individuals and communities in which all needs are (what would be considered) religious.

In spite of these differences between the cases of Christianity and Islam, the aforementioned conceptual categories of religion and science discourse have been applied often and variously to Islam. This has produced a body of literature that attempts to show that while Christianity eventually accepted and even supported the new scientific knowledge that emerged in Europe in the seventeenth century, along with the institutional structures needed to establish science on a solid footing, Islam did not; hence its failure to produce a Scientific Revolution parallel to that of Europe (Huff 1993). Furthermore, it is argued that this reveals something inherently wrong with Islam in this regard—that it abhors innovation, free thinking, and objective inquiry, which are all considered necessary preconditions for science. These studies then attempt to show that since science is produced through innovation and freethinking, and since in Islam "the idea of innovation in general implied impiety if not outright heresy" (Huff 1993, 234), science could not find a home in Islam. Faced with historical evidence of the existence of an eight-hundred-year-long tradition of scientific research in Islamic civilization, these studies then consider it an anomalistic case of the survival of "foreign" (sometimes called "ancient") sciences not because of but despite Islam.

This "Islam versus foreign sciences" thesis was first propounded by the Hungarian Orientalist Ignaz Goldziher (1850–1921) in his paper entitled "The Attitude of Orthodox Islam Toward the 'Ancient Sciences'" (Goldziher 1915), and it has since become a favorite point of departure for building a "conflict model" for Islam and science a la Auguste Comte and a host of other philosophers, including those who conceive religion and science as nonoverlapping domains.

SCIENCE AND ISLAM: THE NATURE OF THE RELATIONSHIP

Using the conceptual categories inherent in Islamic understanding of knowledge—whether scientific or otherwise—we can reformulate the question of the Islam and science nexus. Knowledge is *ilm* in Arabic, a word that frequently occurs in the Qur'an. Knowledge is considered meritorious; *those who know and those who do not know are not equal*, a verse of the Qur'an tells us (Q. 39:9). The Prophet of Islam said that "scholars are the inheritors of the Prophets." He also advised Muslims to "seek knowledge from the cradle to the grave." The acquisition of knowledge is virtuous; it ennobles humanity and it serves its needs. In the case of individuals, a certain amount of knowledge of Islam is deemed essential; for a community, knowledge of various sciences is essential for fulfilling the practical needs of the community. This recognition has produced two categories of obligations: personal and communal. It is the personal obligation (*fard 'ayn*) of a believer to have a certain amount of knowledge of his or her *dîn*, but it is not everyone's obligation to have expertise in astronomy or mathematics; this is instead the obligation of a community, if the need exists. Thus defined, scientific knowledge, whether furthering our understanding of the cosmos and its working or merely fulfilling the practical needs of the community, becomes a "religious" duty incumbent on the whole community, meaning thereby that a certain number of individuals from the community must pursue it with the full financial, logistic, and moral support of the entire community. It is this religious obligation that provides a nexus between Islam and the quest for scientific knowledge.

The conceptual scheme for the "interaction of science and religion" that emerges from this primary understanding of the nature and function of knowledge removes the duality inherent in the two-entity model, and allows us to understand the scientific endeavors of Muslim scientists and scholars of the classical period on their own terms. "I confined this book," wrote al-Khwarizmi in the introduction to his *Algebra*, "to what men constantly need to calculate their inheritance and legacies, [their] portions and judgments, in their trade and in all their dealings with one another [in matters involving] measurement of land, the digging of canals, and geometrical [calculations], and other matters involving their crafts" (Khwarizmi 1989, 4). In writing his book, which would inaugurate the science of algebra, al-Khwarizmi was fulfilling a *fard 'ayn*, for which (he wrote) he hoped to receive recompense from the Creator.

It can be argued that perhaps not all Muslim scientists saw their scientific research in this manner; that they were interested in science for its own sake, or that they were merely pursuing a career, providing bread and butter for their families. While these arguments hold some weight, and while it may even be shown that some Muslim scientists of the period under consideration had no or very little commitment to Islam, these and similar arguments do not render invalid the aforementioned Islamic conceptual framework of knowledge and its pursuit. The two categories mentioned above (personal obligations, *fard 'ayn*, and communal obligations, *fard kifayah*) are Islamic legal terms deeply entrenched in Islamic beliefs and practices.

It is essential to reformulate the questions related to the relationship between Islam and the Islamic scientific tradition in terms that do not impose foreign conceptual categories. Science is a civilizational activity; it fulfills the needs of a given civilization by providing reliable and verifiable knowledge about the physical world. It is pursued by men and women whose understanding of the physical world they explore is directly related to their belief system and worldview. The way a scientist understands the origin and working of the physical world is extremely important to his or her approach to it. It is this understanding of the origin and working of the physical world that forms the matrix from which emerges the relationship between the scientist and science as well as between the scientist and his or her way of being (*dîn*).

Seen in this way, the Islam and science nexus becomes a set of inherent and organic relationships between science, scientists, and their beliefs and practices. In a way, it is the nexus between what an individual perceives as his or her personal obligation (*fard 'ayn*) and his or her role in fulfilling a communal obligation (*fard kifayah*). This does not suggest by any means that there were no tensions or conflicts within the Islamic scientific tradition. All that is being suggested is that Islam views all knowledge—whether scientific or otherwise—through its own unique perspective in which there is a certain unity of knowledge, a certain direction, and a certain purpose. The methodology being proposed here explores the Islam and science nexus as a set of dynamic relationships that arose out of the particular Islamic concept of knowledge (in this case, scientific knowledge) and its function, the needs of the community, the role of individual scientists who deemed it their duty to fulfill these needs, and the natural instinct to acquire knowledge, which has always been the main driving force for exploring the world of nature. As al-Biruni tells us,

> I say, further, that man's instinct for knowledge has constantly urged him to probe the secrets of the unknown, and to explore in advance what his future conditions may be, so that he can take the necessary precautions to ward off with fortitude the dangers and mishaps that may beset him. (al-Biruni 1967, 5)

A possible rebuttal to this methodology would be the presence of non-Muslim scientists and scholars within this scientific tradition. It is true that many non-Muslims participated in the making of the Islamic scientific tradition, especially in its early phase, but on closer examination it becomes obvious that the scientific enterprise cultivated in the Islamic civilization cannot be divided into Muslim and non-Muslim categories; it was a tradition that explored the world of nature from within the overall worldview provided by Islam and, as such, even non-Muslim scientists and translators who participated in this enterprise worked within that overall framework. This should not seem odd. After all, the thousands of Muslim scientists now working in the modern enterprise of science built upon a worldview other than that of Islam have not made this enterprise "Islamic."

A final point on methodology pertains to the role of revelation in Islam. The Islamic concept of knowledge is ultimately linked to revelation, which is considered the only absolutely real and true source of knowledge. In our context, this is to be understood in the sense that whatever knowledge one gleans from or of the physical world by one's external physical senses (touch, smell, taste, sight, and hearing, or by scientific instruments that are extensions of these senses), must be processed in the light of revealed knowledge. Revealed knowledge outlines a certain order of things and their relations. Knowledge derived from senses or their extensions is examined in the light of this order. A star or a moon does not exist by itself, or for itself; it exists within a vast universe populated by numerous other things and, as such, in addition to its own existence as a thing, it has an existence *in relation* to other things. This relational existence provides the framework in which its own existence is examined. The evening star that rises over Samarqand at a certain place and time during the summer months has its own existence, but an al-Biruni or an al-Khwarizmi studying its rising and setting by making observations is using this data for constructing a model of the universe in which the evening star is but one entity. This model of the universe has, broadly speaking, an Islamic framework, and the integration of the scientific data on this star into the greater universe is what is meant by "processing" the data in the light of revelation. The scientist is also making these observations and measurements for a purpose other than, and in addition to, advancing knowledge about the evening star; he is a human being existing within a society that has certain needs that he, by dint of his education, training, resources, and personal preferences, has taken upon himself to fulfill, hoping that "the learned would reward [his] endeavor," as al-Khwarizmi said, "obtaining for [him] through their prayers the excellence of Divine Mercy" (al-Khwarizmi 1989, 66).

BASIC ELEMENTS OF THE ISLAM AND SCIENCE NEXUS

Contemporary religion and science discourse is broadly concerned with three basic themes:

Consonance, Dissonance, Neutrality: Is a certain religious tradition supportive of science? Does it oppose it? Is it neutral? What has historically been the nature of this relationship?

Origins: In what ways does a given religion understand the nature of cosmological and biological origins? How does this understanding produce consonance and/or dissonance with scientific perspectives on origins?

Moral and Ethical Issues: (i) those dealing with the impact of modern science and technologies on the planet, its resources, and the environment; and (ii) certain new issues that have arisen due to the development of new technologies, especially in the biomedical sciences, including those arising out of technologically assisted parenthood, surrogate motherhood, and genetics.

The first set of questions arises out of the two-entity model already discussed in detail. Since the two-entity model is inapplicable to Islam and science (as explored in earlier chapters), there is little to discuss under this heading. Yet, since a large number of secondary works have consistently advocated the "Islam against science" doctrine, it is necessary to clear some basic misconceptions. The question on origins has two aspects: cosmological origins and biological origins; the latter includes the theory of evolution proposed by Darwin and its variations proposed by his successors. In this chapter we only discuss cosmological origins; the question of biological origins is discussed in Chapter 7.

Since in this chapter we only explore the nexus between Islam and science during the eighth to the sixteenth centuries, the third set of issues is irrelevant here, for these technologies did not exist during that time. We will, however, discuss these while dealing with the relationship between Islam and modern science.

STORM IN A CUP OF TEA: ISLAM VS. SCIENCE OR ISLAM FOR SCIENCE?

As far as science is concerned, Islam is definitely a hurdle in its propagation; it can be said that there is something inherently wrong with Islam that does not allow science to flourish. This is why, in spite of the enormous oil wealth available in the Middle East, no Muslim country is producing science today. Historically speaking, it is true that a large part of Greek scientific texts was translated into Arabic and made available to Muslims, but even this did not produce any original science, and certainly not the kind that emerged in Europe at the time of Scientific Revolution. It appears that Greek science survived in the Islamic civilization not because of Islam, but despite it. Treated as "foreign sciences," the Greek heritage was always looked upon by Muslim religious authorities with suspicion and hostility and as soon as they could, they destroyed it. Al-Ghazali (d. 1111) was the man most responsible for this. Despite this, a few brilliant philosopher-scientists and physicians need to be mentioned, for, in spite of vigorous opposition, they left behind a small body of work—mainly commentaries on Aristotle's natural philosophy—that had a significant

Ibn Sina's tomb in Hamadan, Iran. Ibn Sina's influence remained strong in the Muslim world as well as in the West for many centuries. © Al-Qalam Publishing.

impact on Western thought when it was translated into Latin. Among these philosopher-scientists, al-Kindi, al-Farabi (d. *ca*. 950), Ibn Sina, Ibn Bajja (d. 1139), and Ibn Rushd are noteworthy.

This is how a very large number of general books on science and religion, as well as those dealing with the history of science, depict the eight hundred years of scientific activity in Islamic civilization. Most accounts actually reduce this time period to half its length by a summary death sentence, which turns this tradition to an inert mass some time in the twelfth century. This is the prevalent view of nonspecialists, who have never touched a real manuscript with their hands and who have never looked at an Islamic scientific instrument of surpassing aesthetic quality and dazzling details, displaying a mastery of complex mathematical theorems. The extent of the entrenchment of this view makes it almost an obligation of anyone writing a new work on Islam and science to first examine evidence supporting this view. When one makes that attempt one finds that all roads lead to Ignaz Goldziher, the godfather of the "Islam versus foreign sciences" doctrine, who first enshrined it in a German paper called "Stellung der alten islamischen Orthodoxie zu den antiken Wissenschaften" (Goldziher 1915). It was translated into English in 1981 by Merlin Swartz under an inaccurate title, "The Attitude of Orthodox Islam Toward the 'Ancient Sciences'" (Swartz 1981). The English title is misleading, as Dimitri Gutas has pointed out, because "it omits the word 'old' ('alte') from the original title, eliminating even this minimal differentiation among the various epochs of Islamic history...omitting the word 'old' in the English context makes all 'orthodox Islam' appear opposed to the study of the ancient sciences" (Gutas 1998, 168).

Among the 155 references cited by Goldziher, however, there is not a single reference to a scientist complaining about the "opposition of Islamic orthodoxy" to his work a la Galileo. This glaring absence of internal evidence has never been mentioned by any critique of Goldziher's position, although there exist at least four important criticisms that have somewhat blunted its influence in recent years (Sabra 1987; Makdisi 1991; Berggren 1996; Gutas 1998).

What is fundamentally problematic in Goldziher's construction is his conception of "orthodoxy" in Islam, as George Makdisi has pointed out:

The use of the term "orthodoxy" implies the possibility of distinguishing between what is true and what is false. This term implies the existence of an absolute norm as well as an authority which has the power to excommunicate those whose doctrines are found to be false or heretical. Such an authority exists in Christianity, in its councils and synods. It does not exist in Islam. (Makdisi 1991, 251)

By positing his "old orthodoxy" against science, Goldziher wanted to contrast it "to some 'new' orthodoxy, and this is identified as Islam in Goldziher's day, which he mentions in the very last sentence [of his paper]" (Gutas 1998): "Orthodox Islam in its modern development offers no opposition to the study of the ancient sciences, nor does it see an antithesis between itself and them" (Goldziher 1915; Swartz 1981, 209). Gutas has noted that this "statement points to the source of Goldziher's rationalistic and even political bias" and he has suggested that Goldziher's hypothesis should be seen in the light of his well-known anti-Hanbali bias.

Goldziher's attitude toward Islam was formulated in the background of the colonization of the Muslim world by European powers that had, in turn, presented Islam as a spent force that could only be derided and vilified. This bias against Islam, which had penetrated all spheres of thought and imagination of European life in the nineteenth century, must have contributed a great deal to the making of Goldziher's intellectual position toward Islam. He was a direct heir to the medieval history of hostility toward Islam. Islam was then studied in Europe not as a true religion but as an invention of Muhammad: many works included in their titles the term *Muhammadanism*, which Goldziher used in the title of his major work *Muhammedenische studien* (1888).

The European attitude toward Islamic science during Goldziher's time can also be judged from an interesting encounter between Jamal al-Din al-Afghani (1838/9–1897), one of the most influential Muslim intellectuals of the nineteenth century, and Ernest Renan (1823–92), the French philologist, historian, and critic. The details of this encounter also show us the complex psychological makeup of Muslim intellectuals in that fateful century during which the European powers colonized almost the entire Muslim world. They also reveal certain aspects of changes in the European outlook on religion and European understanding of the relationship between religion and science at that time.

Religion was then seen as an inhibitor of science. This was first seen in reference to Christianity, but soon this initial recasting of the role of Christianity in Europe was enlarged to include all religions, Islam being particularly chosen for its perceived hostility toward rational inquiry. The idea that Islam was inherently against science was thus nourished under specific intellectual circumstances then prevalent in Europe, and it was in this general intellectual background that the first echoes of the "Islam against science" doctrine are heard.

The encounter between al-Afghani and Renan was based on a public lecture on "Islam and Science" delivered by Renan at the Sorbonne; it was later published in the *Journal des Débats* on March 29, 1883. In his lecture, Renan forcefully repeated the claim (already in the air at that time) that early Islam and the Arabs who professed it were hostile to the scientific and philosophic spirit, and that science and philosophy had entered the Islamic world only from non-Arab sources (Keddie 1972, 189). Al-Afghani, who happened to be in Paris at that time, responded to Renan. His response was published in the same journal on May 18, 1883. In his response, al-Afghani asked rhetorically: "How does the Muslim religion differ on this point from other religions? All religions are intolerant, each in its own way" (Keddie 1968, 182–83). He goes on to accept Renan's hypothesis, but only in general terms:

Whenever the religion will have an upper hand, it will eliminate philosophy; and the contrary happens when it is philosophy that reigns as sovereign mistress. So long as humanity exists, the struggle will not cease between dogma and free investigation, between religion and philosophy; a desperate struggle in which, I fear, the triumph will not be for free thought, because the masses dislike reason, and its teachings are only understood by some intelligences of the elite, and because, also, science, however beautiful it is, does not completely satisfy humanity, which thirsts for the ideal and which likes to exist in dark and distant regions that the philosophers and scholars can neither perceive nor explore. (Keddie 1968, 187)

Renan's condescending rejoinder to al-Afghani, published in the *Journal des Débats* on May 19, 1883, stated that "there was nothing more instructive than studying the ideas of an enlightened Asiatic in their original and sincere form" (Keddie 1972, 196). He found in them a rationalism that gave him hope that "if religions divide men, Reason brings them together; and there is only one Reason." He then reiterated his racialist views, even in praising al-Afghani: "Sheikh Jemmal-Eddin is an Afghan entirely divorced from the prejudices of Islam; he belongs to those energetic races of Iran, near India, where the Aryan spirit lives still energetically under the superficial layer of official Islam." Renan admits "he may have appeared unjust to the Sheikh" in singling out Islam for his attack: "Christianity in this respect is not superior to Islam. This is beyond doubt. Galileo was no better treated by Catholicism than Averroes by Islam." Renan concludes his rejoinder by stating that al-Afghani had "brought considerable arguments for *his* fundamental theses: during the first half of its existence Islam did not stop the scientific movement from existing in Muslim lands; in the second half, it stifled in its breast the scientific movement, and that to its grief" (Keddie 1972, 197).

This is the immediate background to twentieth-century Western literature on Islam and science. Renan was pushed out of the picture, but Goldziher still reigned supreme in this discourse, which construes Islam inherently incapable of producing science. The fatal division on which Goldziher construed his thesis divides knowledge into 'sciences of the ancients' (meaning all works translated from Greek) and 'the sciences of the Arabs' or the 'new sciences'. By ancient sciences he means "the entire range of propaedeutical, physical and metaphysical sciences of the Greek encyclopedia, as well as the branches of mathematics, philosophy, natural science, medicine, astronomy, the theory of music and others" (Goldziher 1915, 185), although he acknowledges the extensive interest "that these sciences aroused from the second century AH on[ward] in religious circles loyal to Islam (and encouraged also by the Abbasid caliphs)." He states that "strict orthodoxy always looked with some mistrust on those who would abandon the science of Shafi and Malik, and elevate the opinion of Empedocles to the level of law in Islam" (Goldziher 1915, 185–86). Thus the entire range of Greek, Persian, Hindu, and other pre-Islamic works are pitted against "pure Islamic sciences." This position has been previously examined in detail, and hence a quote here from that work will suffice:

But when one examines the data used by Goldziher to construct his battle lines, one realizes that these battle lines are boundaries drawn on sand with a clear and pre-conceived purpose which is none other than a specific interpretation of the whole intellectual tradition of Islam. In order to support his various claims, Goldziher had to rely on exceptions, rather than norms, and on fatal distortions of the data by situating the quoted passages in *his* context, rather than in their proper historical context. For example, he states, as proof for his assertion, that "the pious Muslim was expected to avoid these sciences with great care because they were dangerous to his faith", because the Prophet had prayed to God for protection against a 'useless science'. Goldziher states that this *Hadith* of the Prophet "was quoted frequently". In the footnote to this statement, where one would expect to find references to the 'frequent quotations', one finds only a note stating that the *Hadith* is to be found in Muslim (V, 307) and not in Bukhari, but that "it appears with *special force* in the *Musnad* of Ahmad, VI, p. 318". What does it mean for a tradition of the Prophet to appear in the *Musnad* of Ahmad with *special force*? The *Musnad* of Ahmad, like all such works, is a collection of sayings and description of various acts of the Prophet of Islam, arranged according to the narrator, systematically and in a uniform manner without any special treatment reserved for one *Hadith* and withheld from another.... Furthermore, in the text of the *Hadith*, the word used is *'ilm*, which does not mean sciences of the type Goldziher is referring to; *'ilm* means knowledge in general and taken within the context of the Prophetic supplications, it is extremely unlikely that he would be referring to the "foreign sciences." (Iqbal 2002, 78)

INTERNAL EVIDENCE

From the foregoing it should be apparent that the three possibilities normally explored in contemporary religion and science discourse (consonance, dissonance, neutrality) are inapplicable to the case of the Islam and science nexus for the period under discussion. We have a very large number of practicing scientists who are also religious scholars, or have enough grounding in religious sciences to know what was permissible under the Law and what was not. An Ibn Sina, an al-Biruni, or an Ibn Rushd could, therefore, easily challenge any half-trained Mulla who might object to their science on religious grounds. The tone of authority and confidence displayed by these scientists when writing about the religious basis of their science provides an internal evidence against any "Islam against science" doctrine that attempts to show that religious orthodoxy had an upper hand in such matters. This is clearly not the case, as evident from what al-Biruni wrote in the introduction to his treatise on the shadows:

> I say firstly, that the subject of this investigation can hardly be comprehended except after encompassing (knowledge of) the constitution of the universe according to what is shown by demonstration, excluding what the various groups of people apply to it of what they have heard from their ancestors, as well as recourse from the sects to their beliefs, and (also) after (attaining) the capability of dealing with its varying situations, in which one cannot dispense with arithmetic and deep investigation of it by geometry.
>
> Verily, (even) he who has studied much in the sacred books may not be separated from the mass of the common people, nor from their conviction that this art is contradictory to religion, contrary to divine (Muslim) law; that it is a forbidden pursuit, and an abrogated and forsaken practise. Nothing impels him to this belief but his ignorance of what impugns religion so that he might (properly) support it, his revulsion from the unfamiliar which he inherits from [his likes] before him, and his inability to distinguish what is (truly impugning to religion) from what is not. (al-Biruni tr. 1976, 6)

It is, thus, safe to say that no so-called Islamic orthodoxy could have opposed the practice of science by scientists who were themselves eminently qualified to discern what was their religion's position on various aspects of their chosen fields of research. Not only were many Muslim scientists of the period deeply rooted in the religious tradition, they were qualified enough to write books on the same "Islamic sciences" that Goldziherism poses against their natural science.

Some contemporary projections of "Islam against science" doctrine confound the issues entirely: philosophy is linked to natural philosophy and religion to theology and then, in one sentence, all of these distinct fields of study are uniformly applied to provide "examples" of Islam's opposition to science. Most of these accounts are by authors who base their opinions on questionable translations and secondary sources that repeat a stock account first formulated in the nineteenth century. These works display little understanding of the specific nature of Islamic philosophy (*falsafah*), which had various strands quite distinct from Greek philosophy, including the profound tradition of *hikmah* (wisdom) philosophy discussed in more detail in the following section. Another practice in these confused accounts is to use debates internal to the religious sciences—such as debates on the role of reason in interpretation of revealed knowledge—to build a case for the "Islam against science" doctrine. These debates are often taken out of their proper context, and their carefully chosen terminology (specific to the disciplines in which these debates fall) is disregarded. As an example we will discuss a single case of this type.

"Persecution and harassment of those who advocated the use of reason to explicate revelation are unknown in the medieval Latin West after the mid-twelfth century," we read in one such work:

> How different it was in Islam, if we judge by a question that Ibn Rushd (Averroes) posed in the twelfth century in his treatise *On the Harmony of Religion and Philosophy*. In this treatise, Ibn Rushd sought to determine "whether the study of philosophy and logic is allowed by the [Islamic] Law, or prohibited, or commended—either by way of recommendation or as obligatory" (Averoes 1976, 44). In the thirteenth century, Ibn as-Salah ash-Shahrazuri, an expert on the tradition of Islam . . . issued a written reply (*fatwa*) to a question that asked, in Ignaz Goldziher's words, "whether, from the point of view of religious law, it was permissible to study or teach philosophy and logic and further, whether it was permissible to employ the terminology of logic in the elaboration of religious law, and whether political authorities ought to move against a public teacher who used his position to discourse on philosophy and write about it." (Goldziher 1981, 205)
>
> What is remarkable in all this is the fact that, in the twelfth century, Ibn Rushd and, in the thirteenth century, Ibn as-Salah were grappling with the question of whether, from the standpoint of the religious law, it was legitimate to study science, logic, and natural philosophy, even though these disciplines had been readily available in Islam since the ninth century. Ibn Rushd felt compelled to justify their study, while Ibn as-Salah, astonishingly, denied their legitimacy (as we saw earlier in the chapter). *I know of no analogous discussion in the late Latin Middle Ages in which any natural philosopher or theologian felt compelled to determine whether the Bible permitted the study of secular subjects. It was simply assumed that it did.* (Grant 2004, 241–42, emphasis added)

Disregarding the presence of Goldziher, two noteworthy aspects of this passage are (i) the author's unstated aim to present a comparison between Christianity and Islam (the italicized portion of the quotation) in which he attempts to show two highly respected Muslim scholars (Ibn Rushd and Ibn as-Salah) mired in debate on an issue that had long been resolved in Christianity and (ii) his lack of understanding of the technical terms used in the original texts. Ibn Rushd's celebrated treatise, *Kitâb faṡl li'l-maqâl wa taqrîr mâ bayna al-sharî'ah wa'l hikmah min'l ittišâl*, mistranslated as *On the Harmony of Religion and Philosophy*, has nothing to do with science and religion debates whatsoever. Grant is apparently using George Hourani's English translation (listed in his bibliography), and not the original

Arabic, but even this problematic translation with an incorrect title clearly indicates that the subject of the treatise is to determine, from an Islamic legal perspective, the nature, limits, and conditions of the use of *falsafah* (philosophy) and *mantiq* (logic). Ibn Rushd, let us recall, was born into a family of distinguished scholars and jurists who had held the office of Grand *Qâdi* (Judge) for two generations before his birth; he himself was to become the preeminent Grand *Qâdi* (Judge) of Cordoba and was duly trained in *sharî'ah* (Islamic Law) as well as all branches of Islamic learning of the time, including jurisprudence (*fiqh*), medicine, and *falsafah*. His works include books on a wide range of topics, including philosophy, medicine, Islamic Law, astronomy, and music. The purpose of this particular book, whose Arabic title can be translated as *A Decisive Book and Commentary on What Is Common in Islamic Law and Wisdom*, is to ascertain, as Ibn Rushd himself tells us, "from an Islamic Legal point of view (*ala jahtan nazar al-shari'*) the position of *falsafah* (philosophy) and *mantiq* (logic), whether they as branches of knowledge (*'ulum*) are praiseworthy (*mubah*), prohibited (*mahzur*), or commanded (*ma'mur*)" (Ibn Rushd 1959, 1). To paraphrase it as "grappling with the question of whether, from the standpoint of the religious law, it was legitimate to study science, logic, and natural philosophy" is to read one's own predetermined agenda into a medieval text that was concerned with categories strictly used in Islamic jurisprudence (*mubah, mahzur, ma'mur, mandub,* and *wajib*), categories that cannot be transplanted from their field without doing them gross injustice. Most important, there is no mention of science in the original text! Of course, one can stretch the argument to say that logic is a necessary prerequisite for scientific investigation, but we must remember we are dealing with a medieval text on Islamic Law. At the time Ibn Rushd wrote his treatise, science had been practiced in Islamic civilization for almost four centuries and various branches of science were well-known by their own names (astronomy, alchemy, geography, etc). Ibn Rushd's treatise certainly does not refer to those sciences.

The very pointed indicator to the subject matter of *The Decisive Treatise* (*Fasl al-maqal*) is included in its full title: *ittišal*, from the root T-Š-L, meaning junction and parentage. The purpose of the book has been clearly elucidated by Roger Arnaldez as follows:

What "parentage" (*ittisal*) is there between Islamic religious law (*shariah*) and wisdom (*hikma*)? That is the question discussed in *The Decisive Treatise*. Let us note the expressions used. Averroes is not speaking about the relationship between faith and reason, or between philosophical truth and dogmatic belief: those are general questions which should be examined under the purview of a single, specific form of research, since a relationship can only exist between works of the same kind. This is why Averroes uses the word "parentage", which has a meaning that is more ontological than logical. For him it is actually not a case of bringing a rational view of things into harmony with a religious view, but of discovering whether or not there is a subjective parentage between the way of life according to the wisdom that philosophy has as its goal, and the way of life according to Religious Law, which is revealed. So it is not from the perspective of an abstract problem that Averroes views the issue, but from the concrete perspective of men who are to live and act in this world. They undoubtedly have a practical mind with which they are able to deliberate and to make decisions. But do they use it, and, we might add, are they capable of using it well. Averroes's fundamental idea, which is undoubtedly based on daily experience, is that such is not the case. Religious law is thus in his eyes something that comes to men as an aid to their failing reasons. What remains to be shown is that in acting in accord with the law, they behave according to reason, even though it is not reason that inspires them. (Arnaldez 1998, 79–80)

The real issues discussed by Ibn Rushd in this treatise have nothing to do with Islam and science discourse. What he is interested in doing is to examine whether it is permitted, forbidden, commanded, recommended, or, finally, necessary, to look at the Law with a philosophical or logical eye. Setting the necessary aside, since it belongs to the domain of the rational faculty, the prescribed, the forbidden, the recommended (with its counterpart, the discouraged) and the licit are juridical categories (*akhâm*), in the name of which Muslim jurists seek the nature of laws and qualify the acts that depend on them. Consequently, Averroes is not going to examine the Law on the basis of reason, but reason on the basis of what characterizes laws. (Arnaldez 2000, 80)

Before closing this section, let us note that the tension that existed in the Islamic tradition as a result of the arrival of Greek philosophy needs to be examined in its proper context in a work on philosophy and religion, for the debates that originated from this tension were philosophical debates and had little impact on the natural sciences. It is also important to understand that certain philosophers had political ambitions, were appointed to ministerial posts, and were, therefore, part of palace intrigues that sometimes led to their persecution. These instances have to be examined within the social and political context of the Abbasid Empire (or the specific Sultanate where the incident took place); these cannot rightfully be cited as persecution of philosophers because of their philosophy, as is often done. In one such work a rather dramatic account of the persecution of al-Kindi is made out to be a general situation prevailing in Islamic civilization:

As we have already mentioned, al-Kindi, al-Razi, Ibn Sina, and Ibn Rushd were among the greatest philosophers. All were persecuted to some extent.
Al-Kindi's case reveals important aspects of intellectual life in Islam. The first of the Islamic commentators on Aristotle, al-Kindi was at first favorably received by two caliphs (al-Mamun and al-Mutassim), but his luck ran out with al-Mutawaakil, the Sunni caliph mentioned earlier. According to Pervez Hoodbhoy, "It was not

hard for the *ulema* [religious scholars] to convince the ruler that the philosopher had very dangerous beliefs. Mutawwakil soon ordered the confiscation of the scholar's personal library.... But that was not enough. The sixty-year-old Muslim philosopher also received fifty lashes before a large crowd which had assembled. Observers who recorded the event say the crowd roared approval with each stroke" (Hoodbhoy 1991, 111). The other four [sic] scholars were also subjected to some degree of persecution, and a number of them had to flee for their safety. (Grant 2004, 239–40)

This highly inaccurate account of intellectual life in Islam is based on a book written by Hoodbhoy, a Pakistani physicist who wrote his work of questionable accuracy as a reaction to a superficial movement of Islamization patronized by a military general who had usurped power in a midnight coup. Hoodbhoy based his account on another secondary work, *The Genius of Arab Civilization* (Hoodbhoy 1991, 111). None of these lead us to any primary source that can testify to the public flogging. Grant adds certain details not found in Hoodbhoy.

What we actually know, from the four classical primary sources (Ibn Nadim, al-Qifti, Ibn Nabatah, Ibn Abi Usaibiah) that provide the bulk of known information about Muslim scholars, is rather different. While in Baghdad, al-Kindi was appointed tutor to Ahmad, the son of the Caliph al-Mu'tasim, and was highly respected. Ibn Abi Usaibiah's important work *Tabaqat al-Atibba'* tells us of al-Kindi's great fame, his advanced knowledge, his famous library, and also about the favors he received from the caliphs (including al-Mutawakkil, the "Sunni" caliph—the other caliphs were also Sunni—made out as the villain of the case). The full account of the "public beating" narrated by Ibn Abi Usaibiah does not mention *ulema* at all, instead he names two contemporary rivals: Muhammad and Ahmad, the sons of Musa ibn Shakir, who lived during the reign of al-Mutawakkil, were conspiring against everyone who was advanced in knowledge. They sent a certain Sanad ibn Ali to Baghdad so that he might get al-Kindi away from al-Mutawakkil. Their conspiracies succeeded to the point that al-Mutawakkil ordered al-Kindi to be beaten. His whole library was confiscated and put in a separate place, labeled as the "Kindian Library" (Sharif 1963, vol. 1, 422).

In short, it is safe to say that secondary literature on Islam and science is full of numerous inaccuracies that beget more inaccuracies in a vicious cycle. It is time, however, to turn to one of the most important issues in the discourse: beginnings.

IN THE BEGINNING

Nothing is more important for religion and science discourse than the question of origins. This is the most important issue in this discourse, not merely due to Darwinian and neo-Darwinian accounts of origins but because the questions related to the origin of the cosmos and life have a direct impact on all other issues of the discourse. How and when did the cosmos come into existence? Was it created by a Creator—as religious traditions tell us—or did it come into existence on its own from some eternal matter that existed for itself? If God created it, out of what was it created? And how did that *thing* out of which the cosmos was created come into existence? If the cosmos is believed to have come into existence as an act of a Creator who specifically chose to create, then our understanding of the created cosmos is directly related to the Creator, the purpose of creation, and the nature of the God–humanity relationship. If, on the other hand, the world is considered to be a self-emergent product of a big or small bang, or that of an autonomous wave of energy floating in an eternal universe billions of years ago, and if life is a derivative of the cooling, splitting, and re-formation of inorganic matter resulting from the former process, then our entire conception of human existence on Earth is conditioned differently. Thus, the question of how the cosmos came into existence is for the religion and science discourse what a nucleus is for an atom, and all other issues in the discourse are like electrons that revolve around that nucleus. The question of origins has dominated the science and religion discourse since antiquity and it continues to hold a special position in the contemporary discourse.

The question of origins became a central issue in the Islamic scientific and philosophical traditions after the translation of the Greek texts in the eighth century, and remained so for several centuries. The main question discussed by a host of scientists, philosophers, and religious scholars pertained to the temporal beginning of the cosmos versus its eternity. Muslim philosophers of the first four Islamic centuries, heavily indebted as they were to the Aristotelian tradition, attempted to "Islamize" Aristotle's eternal universe, while those who opposed them considered their position to be a sign of disbelief. These debates have parallels in both the Jewish and the Christian traditions.

In general, the Biblical and the Qur'anic accounts of creation are understood as *creatio ex nihilo*, whereas any kind of eternal emanation is considered a nonreligious position. A closer look at the medieval creation/emanation debates, however, suggests that the point of contention was more refined, focusing on *gratuitous origination* versus *necessity* in the emanation model. In these debates we find Jewish, Christian, and Muslim scholars commenting on each other's positions. For example, Moses ben Maimon (1135–1204), commonly known as Maimonides, and Thomas Aquinas (*ca.* 1225–1274) argued for necessity; the question of creation in time or without an absolute beginning was secondary for them, though they understood the Biblical account of creation as involving an initial moment. In forcefully stating his philosophical position,

Maimonides was simultaneously responding to both Aristotle and Ibn Sina, the most representative follower of Aristotle among Muslim philosophers.

In the Islamic tradition, initial ideas on Origins were formulated solely in the light of Qur'anic data and the sayings of the Prophet. Subsequently, under the influence of Greek thought, various philosophical cosmologies arose. Certain philosophers accepted Aristotle's ideas about the origins and attempted to harmonize them with Qur'anic data; others opposed these efforts. This generated tension that lasted for several centuries. To explore these debates in more detail, let us begin with the Qur'anic data and the tradition of sacred cosmology that arose from reflection on these data.

QUR'ANIC DATA

As mentioned in the introduction, the sciences of the Qur'an were the first to appear in the Islamic intellectual tradition. This was followed by the sciences dealing with the life and sayings of the Prophet of Islam. Thus, when the scientific and philosophical traditions emerged in Islamic civilization they emerged from within a specific intellectual context shaped by the Qur'anic worldview.

The Qur'anic verses on the origins of the cosmos appear in various chapters and are invariably connected with the arguments for the three main themes of the Qur'an: *Tawhid* (Oneness of God), *Risalah* (Prophethood), and *Ma'ad* (Return). These verses also stress the absolute nature of the creative act of God, for when *He desires to create something, He merely says 'Be' and it is* (Q. 36:82). Creation is seen as God's bounteous gift; one of the most recurring themes of the Qur'an is God as Creator. This is reflected in the frequency of variants of the idea of creation, such as *khalaqa, bada'a, fatara*, and *ja'ala*, about which more shall be said in the next section. In addition to the general account of creation, the Qur'an also offers a more specific description, which we have already outlined in Chapter 3: God created the Heavens and the Earth and all that is between them in six days (Q. 7:54-56; 25:59); He created the Earth in two days (Q. 41:9); and *placed therein firm mountains* [towering] *above its surface and bestowed blessings on it, and equitably apportioned its means of sustenance in four days* (Q. 41:10); He turned toward the heaven, which was [yet but] smoke; and He said to it and the Earth: *'come both of you, willingly or unwillingly' to which both responded, 'we do come in obedience.' And He then decreed that it* [the smoke] *become seven heavens in two days, and He imparted unto each heaven its function* (Q. 41:12); He created the seven heavens (Q. 2:29); one upon another (Q. 67:3); *through them flows down from on high, His Will* (Q. 65:12). Moreover, the Qur'an is particularly emphatic that all that exists is sustained by

God, whose attribute *Rabb* means the One Who sustains. It also repeatedly points out that God's creation has a purpose and a plan (Q. 15:21; 25:2; 30:8-9). God has adorned the sky with stars (Q. 67:5) and is the One who has set in motion all the stars and planets so that humanity may be guided by their positions in its travels (Q. 6:97); He is the One Who covers the day with the night and the night with the day (Q. 7:54; 39:5). He has created the day and the night (Q. 21:33), the sun and the moon, each revolving in its precise orbit (Q. 39:5).

THE RADIANT COSMOGRAPHY

Reflections on Qur'anic cosmological data by the Companions of the Prophet were supplemented by his sayings on the creation of the heavens and the earth. This was to give birth to an Islamic cosmological tradition based on the interpretation of the Qur'anic data, the sayings of the Prophet, and reflections on these two primary sources by the Companions of the Prophet. This was later supplemented by scientific observations. During the subsequent centuries, many other cosmological schemes and models were formulated by Muslim scientists and scholars, but all of those schemes were compared to this early cosmology. This first cosmology has been called the Radiant Cosmography (*al-hay'a as-saniya*) by Jalal al-Din as-Suyuti (d. 1505), whose book of the same title summarizes eight centuries of scholarship on this topic.

Four attributes of God mentioned by the Qur'an are especially relevant to this cosmology: *al-Khâliq* '(the Creator), *al-Bâri* (the Maker), *al-Mušawwir* (the Shaper), and *al-Badî'* (the Originator). The early commentators explained these attributes in detail. They also wrote exegeses on the verses related to the creation of the cosmos and produced a small body of literature that outlined essential aspects of this early Islamic cosmology much before the emergence of science in Islamic civilization and before any translations were made from Greek, Indian, or Persian sources. There are two main aspects to this early cosmology: spiritual and physical. The spiritual cosmology came into existence as a result of profound meditation on the cosmological verses of the Qur'an, such as the famous "Verse of the Throne" (Q. 2:255) and the Light Verse (Q. 24:25). This cosmology envisions a spiritual hierarchy of existence and uses symbolic language. The physical cosmos is not absent from this cosmology, but it is perceived in relation to the nonphysical. Some of the most eminent commentators of the Qur'an from the generation of the Companions of the Prophet have left fairly specific comments on the creation of the cosmos. These comments were passed down, generation after generation, were elaborated and explained by other scholars, and have been meticulously preserved in exegetical literature. The twenty-ninth verse of the second chapter of the Qur'an, for instance, mentions the creation of the seven heavens in a succinct manner: *He [God] is the One Who created for you all that is on earth; then He paid attention to the sky and fashioned it as seven perfect* [heavens]; *He has full knowledge of everything* (Q. 2:29). Commenting on this verse, a number of Companions, including Ibn Abbas (d. 687)—the youngest of the four most eminent commentators of the Qur'an from the generation of the Companions—is reported to have said that "God's Throne was on water when nothing other than water had been created yet. When He decided to bring creation into being, He brought out smoke [or, steam] from the water, which raised upwards, and from this He made the skies. Then the water dried up and He made one earth from it; then He fashioned seven earths from this one in two days... likewise, he fashioned seven heavens from the one He had made from the smoke" (Ibn Kathir 1998, 214).

As-Suyuti's aforementioned work deals with a range of cosmological concepts, situating everything within the Qur'anic worldview. Thus we find sections on the Throne verse (Q. 2:255) and the verse of the Tablet and the Pen (Q. 68:1), as well as on the seven heavens and the seven earths, the dimensions of the cosmos, sun, moon, and stars, the comet, the Milky Way, rainbows, the night, the day, hours, and water and winds, mountains and rivers.

It must be kept in mind that the Qur'anic cosmological verses appear in the Qur'an within the context of the Qur'anic message. They have a specific vocabulary and are often cited as proofs of the Oneness of God, His attributes, and His inexhaustible knowledge. Nevertheless, they do refer to the physical cosmos and its creation. This physical cosmology is part of the broader creation theme of the Qur'an, which has multiple meanings that cannot be reduced to the mere physical level.

Based on Qur'anic data, this cosmology became the counterweight to the Aristotelian cosmology that came into Islamic thought in the translation movement. These debates took place in an intellectual atmosphere ripe with enthusiasm, in a social milieu full of new developments and constant internal strife, in an ever-expanding geographical setting, and in places as diverse as mosques, madrassahs, courts, palaces, markets, bookstores, and observatories. Those who took part in these debates were Qur'an commentators, philosophers, scientists, Sufis, and men of letters and learning.

The early exegetical tradition favored creation in time, for eternity was an attribute that could only be applied to God. *Allah, there is no deity but He, the Living, the Everlasting; slumber seizes Him not, nor sleep* (Q. 2:255). This tradition also favored *creatio ex nihilo*, for anything preeternally existing with God would compromise the Oneness so central to the Qur'an. Creation in time is called *hudûth* in Arabic, and anything that is contingent is known as *hâdith*. Eternity is denoted by the word *qidam* and "eternal" by *qadîm*. Built into this technical terminology is the notion of an Absolute, Sovereign God, upon Whom depends all that exists. Inherent to the term *hudûth* is the idea that the thing referred to depends upon something external for its existence, and that it has come into existence after being nonexistent. These terms have slightly different meanings in philosophical, *Kalam*, and Sufi texts, and sometimes various thinkers have used them in different ways, but they in essence describe God's relation to the world in the Radiant Cosmology.

THE MAKING OF THE CONFLICT

The large amount of scientific and philosophical literature brought to the Islamic tradition by the translation movement included specific cosmological schemes of the Greeks, the Persians, and the Indians. These cosmologies were then viewed, examined, and contrasted with the Qur'anic cosmology by Muslim thinkers; this produced a large body of literature within the Islamic philosophical, scientific, and mystical traditions. This process went on for several centuries, producing new and refined schemes. In essence, it was an attempt to find ways to harmonize the received data and cosmological theories with the Islamic concept of cosmos. Although the process was complex and often fraught with tension, it was not a case of "Islam opposing science," for the pre-Islamic cosmologies were anything but scientific and were rather philosophical constructions. Some of the received material was even in harmony or quite close to the Islamic view of the cosmos, though necessarily formulated in different terms. Pythagorean and Platonic schemes, for instance, were of great interest to the School of Illumination (*al-Ishraq*).

The greatest influence upon the Islamic philosophical cosmology came from Aristotle and his school. Translated into Arabic, his *Metaphysica*, *Physica*, and *De caelo* influenced Islamic philosophical cosmology like no other Greek, Persian, or Indian work. His influence especially impacted the Peripatetic School. The Aristotelian corpus was, however, viewed by Muslim thinkers more often than not through the eyes of his Neoplatonic commentators (Nasr 2001, 365). Plotinus (d. 270 CE), the founder of neo-Platonism, and his student Porphyry (d. *ca.* 301 CE) had a significant role to play in the remaking of the Aristotelian corpus. More important for the Arabic reception of Aristotle however were the two sixth-century commentators of Aristotle, John Philoponus (*fl.* first half of the sixth century) and Simplicius (d. after 533); the former was known to the Arabs as Yahya an-Nahwi, John the Grammarian. Philoponus was a Christian neo-Platonist critic of Aristotle, and thus his commentary was especially significant for Islamic tradition, just as it was for Christianity.

Plotinus was most probably an Egyptian by birth. He studied philosophy at Alexandria and taught in Rome, where he settled around 243. *Enneads*, his collected works, contains six books of nine tracts each. Plotinus's description of God comes very close to the Qur'anic description. The transcendence underlined in the Qur'anic verse *There is nothing like unto Him* (Q. 42:11) refers to an All-Encompassing (*al-Muhît*) and All-Knowing (*al-'alîm*) God above and beyond all human conceptions, and who can only be defined *via negativa*, by erasing from the mind any impurity foreign to the idea of pure Divinity. Plotinus's definition *via negativa* quite resembles this Qur'anic description. As opposed to the Qur'anic cosmology, however, he also believed that the world emanates from God by necessity. God thus does not will the world to emerge, because that would imply change in Him. Like Plato, Plotinus believed in the immortality of the soul and in its existence before the body. In his cosmology, the material world was the last emanation, lying below the soul. He considered matter to be an antithesis of the One, having no positive qualities. We will have more to say about the influence of Plotinus on some Muslim philosophers in the next section, but let us mention here that it was his remarkable synthesis of monotheism, philosophical cosmology, and his strong rejection of Aristotle's belief in the eternity of the world that resonated with the views of some Muslim philosophers.

Beginning in the eighth century, Muslim philosophers produced various cosmological doctrines that were often influenced by Greek, Persian, and Indian cosmologies, but that attempted to harmonize the received material with Islamic beliefs. It is in these attempts that we begin to see the influence of Islam on philosophical cosmology. To look into this interaction in more detail, we first restate the essential features of Aristotelian cosmology (having already touched on some of its aspects in Chapter 3), and then proceed with a systematic account of the various cosmological schemes produced by Muslim thinkers between the eighth and the sixteenth centuries.

ARISTOTELIAN COSMOLOGY

Grant has thus summarized Aristotle's cosmology:

For Aristotle, the cosmos was a gigantic spherical plenum that had neither a beginning nor would ever have an end. Everything in existence lies within that sphere; nothing exists, or can possibly exist, outside of it: neither matter, nor empty space, nor time, nor place. Aristotle regarded it as nonsensical to inquire about extracosmic existence, consequently rejecting the possibility that other worlds might exist beyond our own. Within the cosmos, Aristotle distinguished two major divisions: the celestial regions and the terrestrial. The dividing line between the two regions was the concave surface of the lunar sphere. That surface divided two totally dissimilar regions.

The terrestrial region, which lay below the concave lunar surface, was a region of constant change and transformation. It consisted of four elements: earth, water, air, and fire, arranged in this order from the center of the world to the moon's concave surface. All bodies were compounded of combinations of two or more elements. In the terrestrial region, bodies were always coming into being as differing compounds of the four elements, and bodies were always passing away because their elements eventually dissociated to combine with other elements and form new compound bodies. At the center of the universe was the earth, surrounded in many of its parts by water and then air and fire.... In the upper atmosphere of the terrestrial region, just below the concave surface of the moon, Aristotle assumed that comets, shooting stars, and other similar phenomena occurred. He inferred their existence in this region because they were changeable phenomena, and therefore could not occur in the unchanging celestial region. (Grant 2004, 41)

Aristotle attributed the lack of change in the celestial region to a fifth element: celestial ether, an incorruptible, eternal, and noble substance that suffers no change except that of place. It did not come into existence and will never pass away. He regarded the planets and the stars as composed of ether and thus also undergoing only observable change of place. He believed that the celestial region was superior to the terrestrial and had an influence on the latter. His cosmos was an eternal cosmos, without beginning or end. It was "caused" by a final cause, a God eternally absorbed in self-contemplation to the extent that he has no immediate knowledge of his creation. Although not created by God, Aristotle's cosmos was a rationally constructed cosmos in which a certain order prevailed.

In the Aristotelian system all bodies were composites of matter and form. Form is the active and matter the passive principle. A body can change into another through one of the four possible causes: (i) material; (ii) formal; (iii) efficient; and (iv) final-efficient. The material cause refers to the matter from which something is made (e.g., the wood from which a table can be made); the formal cause is the essence or inner structure of a thing (the shape of a table); the efficient cause is the agent or producer of the change (the carpenter who makes the table); the final cause is the end or purpose for which an action is performed (the table in the mind of the carpenter for which he or she has a use). To take another famous example, Aristotle believed that since an acorn has the potentiality to become an oak tree, it will try to realize the form of an oak tree through the operation of the final-efficient cause.

Aristotle believed that these four causes can produce four kinds of changes: (i) substantial; (ii) qualitative; (iii) quantitative; and (iv) change of place. Of these four, the substantial change is the most fundamental, since it involves replacement of one form with another, as when fire reduces a log to ashes. Qualitative change occurs when one of the qualities of a body changes (the color of a leaf from green to yellow). Quantitative change increases or decreases the size of a body without altering its identity. The fourth kind of change moves a body from one place to another.

Aristotle divided all existing things into two kinds: those that exist by nature (such as plants and animals) and those that are products of art (such as tables and chairs). Things that exist by nature have within them a principle of change, whereas products of art have no innate impulse to change. Thus, a tree will change because of its innate nature, whereas a table or a chair will not, unless these products of art are composed of things that do have an impulse to change. Aristotle also regarded nature as a "cause"—a principle of motion and change—that operates for a purpose. It was the function of physics to explore the causes and motions and changes produced by causes.

ISLAMIC PHILOSOPHICAL COSMOLOGIES

In addition to the already discussed "Radiant Cosmology," several other cosmological doctrines appeared between the eighth and the seventeenth centuries. In spite of their roots in Islamic beliefs, they reflect, to a greater or lesser degree, Pythagorean, Hermetic, Greek, Persian, and Indian influences. It is important to note, however, that despite these influences all of these apparently different cosmologies are concerned with reconstructing the cosmos so that Islam's most basic tenet, Oneness of God (*Tawhid*), becomes its operative principle. In cases where this could not be accomplished to a satisfactory degree, conflicts engaged the best minds of that time in debates that have left a rich legacy of intellectual interaction: an al-Ghazali versus an Ibn Sina, an Ibn Rushd against an al-Ghazali. It should again be noted that these conflicts are not, strictly speaking, conflicts between science and Islam; rather, these are conflicts between views of one philosopher and another, both of whom are attempting to produce a systematic account of the cosmos based on rational thought.

One of the most important philosophical cosmology in Islam belongs to the Peripatetic School, best represented by Ibn Sina. Honorifically called al-Shaykh al-Ra'is, the Grand Shaikh, Ibn Sina's prodigious learning is legendary. Born in Afshanah, a village in present-day Uzbekistan not far from Bukhara, in August or September 980, to the governor of that village, Ibn Sina had mastered many traditional sciences and a large part of Greek philosophy before he was twenty. He studied *Isagoge*, Porphyry's introduction to the *Organon* of Aristotle, with a reputed philosopher of the time, al-Natili, and then moved on to study the works of Euclid and Ptolemy. Soon his teacher left Bukhara, but he "continued the study of texts—the original and commentaries—in natural sciences and metaphysics." At the age of sixteen, he embarked upon reading logic and other parts of philosophy, as he tells us in his short autobiography:

Then, for the next year and a half, I dedicated myself to learning and reading; I returned to reading logic and all parts of philosophy. During this time I did not sleep completely through a single night nor devoted myself to anything else by day ... thus I mastered the logical, natural, and mathematical sciences, and I had now reached the science of metaphysics. I read the *Metaphysics* [of Aristotle], but I could not comprehend its contents, and its author's objective remained obscure to me, even when I had gone back and read it forty times and had got to the point where I had memorized it. In spite of this I could not understand it nor its objective and I despaired of myself and said: "This is a book for which there is no way of understanding". But one day in the afternoon when I was at the booksellers' quarter, a seller approached me with a book in his hand which he was selling by calling out loud. He offered it to me but I refused it with disgust believing that there was no merit in this science. But he said to me, "buy it, because its owner needs money and so it is cheap. I will sell it to you for three *dirhams*". So, I bought it and, lo and behold, it was Abu Nasr al-Farabi's book on the objects of *Metaphysics*. I returned home and was quick to read it, and in no time the objects of that book became clear to me because I had got to the point of having memorized it by heart. I rejoiced at this and the next day gave much in alms to the poor in gratitude to God, the Exalted. (Ibn Sina tr. 1974, 25–35)

Ibn Sina was to remain committed to the Aristotelian framework for the rest of his life, and leave behind a legacy of philosophical texts that would influence the course of Islamic philosophical thought for many centuries. Later, he and Ibn Rushd, a great philosopher from Spain committed to a similar understanding of the cosmos, would become the two most important Muslims to influence philosophy in the Latin West.

Ibn Sina outlined his cosmological doctrines in his major philosophical works, which include *al-Shifa* (*The Book of Healing*), *al-Najat* (*The Book of Salvation*), *al-Mabda' wa'l Ma'ad* (*The Beginning and the End*), and the *Isharat* (*The Book of Directives and Remarks*); his great work on medicine, *al-Qanun fi'l tibb* (*Canon of Medicine*), also contains a great deal of cosmological thought.

Ibn Sina envisioned the cosmos in an Aristotelian manner, used his terminology (form, matter, accidents), and defined change as the passage from potency to act. He lists six causes instead of Aristotle's four, but his two additional causes (the matter and form of the composed) are reducible to the given of the accident and the form of the matter, respectively.

Ibn Sina's cosmology, however, differs from Aristotle's in two important respects. For Aristotle, the distinction between essence and existence was only a logical distinction. Ibn Sina made this distinction ontological. That is to say, Ibn Sina considered this distinction present in every being except the Divine Being. This allowed him to preserve a uniquely Islamic view of existence of things, because every existent in which this distinction is to be found must come into existence through an agent in whom these two are united, and this leads to the division of beings into contingent and necessary. The second major difference between the cosmologies of Aristotle and Ibn Sina is the manner in which Ibn Sina attempts to preserve the Oneness of the Necessary Being while at the same time seeking to explain multiplicity. He did this by adopting emanation theory in a manner akin to the neo-Platonists. Both of these distinctions are typically Islamic needs of a philosopher who was otherwise committed to Aristotle like no other Muslim philosopher before him.

In a number of works, such as his *Book of Healing* and *The Beginning and the End*, Ibn Sina first proves that there is a Necessary Being, then establishes the Uniqueness and Oneness of this Necessary Being. It should be noted that Ibn Sina's Necessary Being is not the same as Aristotle's deity. His Necessary Being is the Most Beautiful, Perfect, and Best, a Being prior in existence to everything and the source of the existence of everything. Moreover, it is a Being free from matter, One and Simple in all respects.

In Ibn Sina's cosmology, multiplicity arises through emanation from Pure Being. There are grades and degrees of existence. In this system, Allah is at the summit. He brings into being the Pure Spirit, called the Primary Cause. From this Cause come the souls, bodies of the spheres, and the intelligences. From the tenth intelligence emerges the sublunary region. According to Ibn Sina, this intelligence is caused by intellectual emanation proceeding from God and ending with the human rational soul. Its primary function is to give corporeal form to matter and intellectual form to the rational soul, hence its name: the giver of forms (*wahib al-suwar*). In this scheme, the beings near the periphery of the universe and close to the Primary Body that surrounds the cosmos are closer to the Necessary Being and are purer than those near the earth. Ibn Sina envisions an impetus in all bodies that is a desire to reach perfection that belongs to the Necessary Being.

Later in his life Ibn Sina wrote three visionary recitals that offer a different cosmological perspective from his better-known Peripatetic works. In these recitals, which are now the last chapters of his *Book of Directives and Remarks*, Ibn Sina describes "the Universe as a vast cosmos of symbols through which the initiate seeking Divine Knowledge, or gnosis (*ma'rifah*), must travel. The cosmos, instead of being an exterior object, becomes for the Gnostic an interior reality; he sees all the diversities of Nature reflected in the mirror of his own being" (Nasr 1993, 263).

This spiritual cosmology was later developed by other schools of thought, especially the School of Illumination (*Ishraqi*), whose illustrious founder, Shihab al-Din Suhrawardi (d. 1187), was initially content with Ibn Sina's philosophy, which he had studied in his youth but which he abandoned after a dream in which Aristotle appeared to him "revealing the doctrine later known as 'knowledge by presence' and asserting the superiority of the Ancients and certain of the Sufis over the Peripatetics" (Suhrawardi tr. 1999, xvi).

The Illuminationist philosophy is based on "unveiling." It gives an important epistemological role to intuition. In logic, "Suhrawardi rejected Peripatetic essential definition, arguing that essences could only be known through direct acquaintance" (Suhrawardi tr. 1999, xx). Suhrawardi believed there were a great many more intellects than the ten of Ibn Sina. His work consists of two parts, the first dealing with logic (in three discourses) and the second with the "science of lights" (in five discourses). Instead of the self-evident substance of Aristotle, here we have light as the foundational entity upon which the entire system of Illuminationist philosophy is built. Thus, the basic metaphysical concepts in this system are light, darkness, independence, and dependence. Light is self-evident and it makes other things evident. Suhrawardi identifies four classes of beings: self-subsistent light, which is self-conscious and is the cause of all other classes of beings; accidental light, which includes both physical light and some accidents in immaterial light; dusky substances (which are material bodies), and dark accidents, which include both physical accidents and some accidents of immaterial light. His cosmogony explains how all other beings came into existence from the Light of Lights (God).

Another outstanding scientist-philosopher who has left us a cosmological scheme different from the Peripatetic cosmology is al-Biruni, who was born eight years before Ibn Sina and died fourteen years after him. Born outside (*birun*) present-day Khiva in 973, al-Biruni had a very independent mind that would not accept anything merely on authority. This is evident from his attitude toward Abu Bakr Zakaria al-Razi, often touted as the most "free-thinking" of the major philosophers of Islam. He studied al-Razi's philosophy, even produced a bibliography of the works of the elder scholar, yet condemned him for his errors (Nasr 1993, 109). Al-Biruni's cosmology is also most clearly imbued with the Qur'anic descriptions of the cosmos. Although he lived at a time when the impact of the translation movement was the strongest, he rejected the arguments of the Greek philosophers on the eternity of the world and produced a cosmology based on *ex nihilo* creation. His cosmological doctrines reveal an amazing synthesis of his observations during travels, his readings, and scientific experiments. As Nasr has pointed out, perhaps the most remarkable feature of his cosmology is the qualitative understanding of time (Nasr 1993, 118). He believed that time does not unfold in uniform manner. In other words, the so-called laws of nature have not been operating in the same manner over the entire length of time since creation. Al-Biruni may have reached at this qualitative understanding of time during his travels in India, because it is a central doctrine of the Hindu cosmology. For al-Biruni, time has a beginning and an end. He looks at geological changes over long spans of time as proof of lack of permanence and describes changes in mountains, rivers, deserts, and other apparently stable and permanent-looking features of the Earth.

Al-Biruni's concept of nature is directly relevant to our book, for here we find empirical evidence for the claim that the Qur'anic view of nature previously outlined deeply affected the way nature was perceived and studied by Muslim scientists of the period between the eighth and sixteenth centuries. This understanding was so deeply embedded that no one felt the need to write a book on the relationship between their science and faith, yet the entire framework for the study of nature was based on the Qur'anic view, which is not merely limited to the theme of creation but encompasses a teleology that reminds awakened hearts to observe the manifest signs of God's Wisdom and Design in nature. That there is nothing superfluous in nature is not due to chance, but is a manifestation of Allah's Wisdom and Power—a bountiful expression of His act of measuring out for each creature which it needs. Or, as al-Biruni puts it:

All our praise and encomiums are for Him alone, Who has so shaped life that each creature can sustain itself and live in a measured way, where there is no excess and no want; and for sustenance has made food the principal cause, and through which each body grows on all sides, so that the food, after being digested, helps it (to sustain itself).

God has made plants content themselves with little food, food that is not consumed rapidly... water seeps in and traverses to their roots. Air, heated by the sun, absorbs moisture from their branches, and transports it upwards. Whatever is thus acquired from below is transported to the branches and causes them to grow. And they produce what they are created for, be it the generation of leaves, or flowers, or fruits. (al-Biruni 1989, 3–4)

Humans may find "fault" in nature, but it is really human shortsightedness that construes a function of nature as a "fault," for "it only serves to show that the Creator who had designed something deviating from the general tenor of things is infinitely sublime, beyond everything which we poor sinners may conceive and predicate of Him" (Nasr 1993, 124).

Some time toward the end of the tenth century or early in the eleventh, al-Biruni corresponded with Ibn Sina on various aspects of the physical sciences and Aristotle's ideas. This correspondence, containing eighteen questions and their responses, not only counters the notion Islamic scientific tradition was only the work of a few individuals working in isolation, but also provides insights into the independence of al-Biruni, who calls into question many well-established Aristotelian notions, often using his own observations and experiments as proofs (al-Biruni 2001–6, 91). A case in point is the sixth question in this correspondence, in which al-Biruni objects to the Aristotelian understanding of the process of solidification of water upon cooling. "I have broken many bottles [by freezing them]," he tells Ibn Sina, "and they all break outward, rather than collapse inwardly." What he leads to is the lower density of water in the solid state than in the liquid state, which is why the volume of water increases upon freezing and flasks break outwardly. This was in contrast to Ibn Sina's views (based on Aristotle). This is a remarkable insight into the nature of the water molecule from the tenth century, when modern-day instruments were not available to detect hydrogen bonding.

Al-Biruni's unique contribution to cosmology lies in the way his entire system is interlinked. He provides a detailed description of the sublunary world, gives geographical details of the various regions of the earth, describes the formation of mountains, clouds, dry and wet lands, the seven climes, and numerous minerals, plants, and animals. In his writings on the animal and plant kingdom, his ideas are directly inspired by the Qur'an, just as his overall scheme of human existence and the human relationship with the world. For example, in his description of the senses of hearing and sight in human beings, al-Biruni states: "Sight was made the medium so that [man] traces among the living things the signs and wisdom, and turns from the created things to the Creator" (al-Biruni 1989, 5). Then he cites the verse *We shall show them Our signs on the horizons and within themselves until it will be manifest unto them that it is the truth* (Q. 41:53).

Islamic cosmological schemes continued to explore various aspects of the discipline after the death of al-Biruni and Ibn Sina, though the groundbreaking works of the tenth and the eleventh centuries remained dominant. A special genre of interest is the "Wonders of Creation" tradition, which sought to describe the entire universe. The twelfth and the thirteenth centuries were particularly rich in these encyclopedic writings.

GOD, CREATION, AND THE ETERNITY OF THE WORLD

In contrast to our own era, the existence of God was not a widespread issue during the Middle Ages (at least in the geographical area under consideration). Atheism and agnosticism did not exist in the public sphere. Muslim, Christian, and Jewish philosophers of that time constructed their metaphysics, physics, and cosmology on the understanding that God exists. What was sometimes disputed was the provability of God's existence, for some philosophers believed that such a proof could be offered rationally while others demurred. Those who attempted to provide proofs for the existence of God can be divided into two broad categories: those who demonstrated the existence of God from the premise of the world's eternity and those who based their proofs on the premise of its creation in time. The issue of the eternity of the world and its creation was, therefore, the most central issue in the Muslim and Jewish philosophical traditions during the Middle Ages (Davidson 1987, 1). What was at stake was not merely hermeneutics, as the issue had a whole range of basic problems attached to it, including the relationship between God and the universe on the one hand and God and humanity on the other. Furthermore, it involved such questions as whether God is a necessary or a voluntary cause. The most contended point revolved around the will of the deity. If the world should be eternal, the deity's relationship to the universe would also be eternal.

Since eternity and necessity are, by virtue of an Aristotelian rule, mutually implicative, an eternal relationship is a relationship bound by necessity; and necessity excludes will. The eternity of the world thus would imply that the deity is, as the cause of the universe, bereft of will. A beginning of the world would, by contrast, lead to a deity possessed of will. (Davidson 1987, 1–2)

Though both premises could be used to construct proofs for the existence of God, the deities arrived at through chain of reasoning based on the two premises would be different. In the argument for the existence of God as the prime mover, Aristotle demonstrated that the world is eternal; its eternal existence is caused; and that this cause is the deity. The Platonic procedure favored by the Kalam tradition, on the other hand, takes creation as its indispensable premise. Since the world came into existence after not having been existent, there must be a creator who brought it into existence. The deity shown to exist by either procedure has three distinct qualities: it is a being which is uncaused, incorporeal, and one. Proof based on the premise of eternity does not show volition to be a characteristic of the deity. This lack of volition became a point of contention. Some philosophers did not take either of the two premises into consideration and so avoided altogether the issue of volition in the deity. Another procedure, adopted by Ibn Tufayl (d. 1185), Maimonides, and Thomas Aquinas, was to use both premises to provide proofs for the existence of God. Ibn Tufayl noted in his celebrated philosophical tale *Hayy Ibn Yaqzan* (*Living, Son of the Awake*) that the issue was unresolvable, and that the existence of God would have been conclusively demonstrated only if shown to follow from both the hypothesis of the world's eternity and its creation (Ibn Tufayl 1972, 81–86).

These debates came into the Islamic tradition from Greek sources along with the entire stock of commentaries on Aristotle's original proof for the existence of God. The translation movement also brought into Arabic the entire stock of refutations of Aristotle's arguments, most important the refutation of John Philoponus, who systematically refuted all the arguments for the eternity of the world put together by Aristotle and Proclus (d. ca. 485). These stock proofs and refutations were supplemented by numerous subtle refinements by Muslim and Jewish philosophers over the centuries. One of the most important aspects of the Islamic as well as Jewish traditions is that whereas Aristotle had taken the world as his point of departure, Muslim and Jewish philosophers often take God as their point of departure.

These debates involved a large number of philosophers, who were sometimes also scientists, but there were no means available to them to test their arguments by experiments such as carbon dating; all they had were philosophical tools such as logic. In the course of these debates, a further refinement of the argument led some philosophers to postulate that matter was eternal, but the world in its present form was created in time. Thus, the entire debate can be divided into three categories: (i) arguments for the eternity of both matter and the world; (ii) arguments for the creation of both matter and the world; and (iii) arguments for the eternity of matter but creation of the world.

That it is impossible for generation to take place from nothing was the foundation of Aristotle's argument, but those who opposed eternity attempted to show that it was a feeble foundation. Its ultimate counter-argument rested on the point that what we observe today to be a law of nature cannot be retrojectively assumed for a time hidden from us. Thus, the fact that we never observe a hen except from an egg or an egg except from a hen cannot be extended beyond our observation to the remote past. Many *Kalam* writers, such as Abd al-Jabbar (d. 1024) and al-Juwayni (d. 1085), refuted the eternity of the world on the basis of this argument, which had already been put forward by John Philoponus.

The terminology of these thinkers, however, is uniquely Qur'anic. They showed that the advocates of eternity (*dahriya*) proceed from what is present and perceivable (*shahid*)—the world as it is today—to what is not present and perceivable (*gha'ib*), applying the same laws of nature, and that this is untenable (Davidson 1987, 30). Jabir bin Hayyan had also used the same argument. In addition to this comprehensive argument, advocates of creation also offered individual and specific refutations.

There are three basic proofs for eternity that take the Creator as their point of departure; these have been succinctly put together by Davidson as

(i) no given moment, as against any other, could have suggested itself to the Creator as the proper moment for creating the universe; (ii) the cause of the universe must be unchangeable and could not, therefore, have undertaken the act of creation after having failed to do so; and (iii) the cause of the universe possesses certain eternal attributes and that the existence of the universe is an expression of those attributes; since the attributes are eternal, the universe, which they give rise to, must likewise be eternal. (Davidson 1987, 49)

Muslim philosophers probably received these proofs through Proclus and added various refinements to them. In Ibn Sina's *Shifa'*, for instance, one finds a rhetorical question: "How within [the stretch of] nonexistence could one time be differentiated for [a creator's] not acting and another time for [his] starting to act? How might one time differ from another?" (Davidson 1987, 50). This main argument was supplemented by many variations. One such variation runs as follows: For the Creator to become active after having remained inactive would have to be due to a new and external factor, which in turn would by necessity have to be due to yet another factor, leading to an absurd infinite regress of causes. This was the position of Abd al-Jabbar (d. 1024), al-Baqillani (d. 1013), Ibn Sina, and Ibn Rushd, among others. Interestingly, the formulation of this argument was further refined by opponents of eternity including al-Ghazali, Fakhr al-Din al-Razi, and al-Amidi (d. 1233). They used the term *tipping the scales* to restate this argument. That is, something had to tip the scales (from nonexistence to existence) at the moment of creation for the world to be created.

In response to this argument, advocates of creation pointed out that since time did not exist prior to its creation, *before* and *after* have no meaning. Following al-Ghazali, Fakhr al-Din al-Razi also offers another argument as a response to advocates of eternity: it is in the nature of God's will to choose a particular time for creation; He can will something irrespective of determinant factors; His knowledge determines the appropriate time for creation (Ceylan 1996, 52). Al-Ghazali's other response to the argument was as follows:

With what [argument] would you deny one who says: "The world was temporally created by an eternal will that decreed its existence at the time in which it came to be; that [the preceding] nonexistence continued to the point at which [the world] began; that existence prior to this was not willed and for this reason did not occur; that at the time at which [the world] was created it was willed by the eternal will to be created at that time and for this reason it was created then?" What is there to disallow such a belief and what would render it impossible? (al-Ghazali tr. 2000, 15)

The standard *kalam* proof for creation was offered from accidents. Since accidents are necessary concomitants of bodies and are subject to generation, bodies too must be subject to generation; since the universe is a body, it must therefore have been generated. This proof was already current in the ninth century (it can be found in the work of Abu al-Hudhayl, who died in 849), and remained a standard response to the advocates of eternity for the next several centuries.

Before closing this section dealing with the pre-Mongol period, we need to mention one more cosmology in some detail, because it is the most important mystical understanding of the cosmos developed in Islamic thought—the cosmological scheme of Ibn al-'Arabi (d. 1240).

MYSTICAL ASPECTS OF ISLAMIC COSMOLOGY

Born in Murcia (southeastern Spain) fifty-four years after al-Ghazali's death in 1111, Muhammad bin Ali bin Muhammad Ibn al-'Arabi al-Ta'i al-Hatim was to become a bridge between the pre- and post-Mongol phases of Muslim history and, perhaps more important, between the western and eastern Sufi orders of the Muslim world. He was to leave behind teachings and insights of several generations of Sufis who preceded him in a vast and systematic corpus that provides a unique insight into the mystical cosmos. He was also to leave behind a tapestry of oral tradition, along with a treasure-trove of technical terms and symbols enriched over centuries. To a Muslim world that was soon to suffer a blow from the Mongols he left a definitive statement of Sufi teachings and a full record of Islam's esoteric heritage. His cosmology is important in itself, but its importance becomes even greater when viewed in the light of its impact on subsequent Islamic thought.

His step-son and disciple in Konya, Sadr al-Din al-Qunawi (d. 1274), served to transmit his teachings to the Muslim East. Qunawi, a Persian by birth, himself a scholar of *Hadith* and a Sufi master of high standing, was also well-versed in Peripatetic philosophy, which he tried to harmonize with the mystical teachings of Ibn al-'Arabi. Qunawi thus serves as a link between the Peripatetic and Sufi cosmologies. He initiated a correspondence with Nasir al-Din al-Tusi (d. 1274), the great Shia theologian who revived Ibn Sina's teachings in the thirteenth century, in order to "combine the conclusions derived from logical proofs with those gained by unveiling, opening, and face to face vision of the unseen world" (Chittick 1989, xix). Thus he was the Shaykh, spiritual master, of Qutb al-Din al-Shirazi, the notable commentator on the Ishraqi philosophy of Suhrawardi, and Fakhr al-din al-Iraqi (d. 1289), the great mystical poet, and an intimate friend of Jalal al-Din Rumi (d. 1273), the author of the *Masnawi*. A century later the teachings of Ibn al-'Arabi inspired another great mystical figure, Abd al-Karim al-Jilli (d. *ca.* 1424), who wrote *al-Insan al-Kamil*. Through Ibn al-'Arabi, Rumi in the East and Abu al-Hasan al-Shadhili (d. 1258) in the West emerged as two of the greatest Sufis in history, and their thought influenced a number of mystical cosmologies both in the East and in the West. In the Indian subcontinent, one can find the impact of Ibn al-'Arabi's cosmology on the thought of Shaykh Ahmad Sirhindi (d. 1624) and Shah Wali Allah of Delhi (d. 1762).

Before we outline Ibn al-'Arabi's cosmology, something must be said about him, for this provides a background for understanding the unique world from which he emerged. He was a member of a distinguished family known for its piety. His father, Ali ibn al-'Arabi, was a man of standing and influence. The famous philosopher Ibn Rushd was one of his friends and Ibn al-'Arabi met him while he was still a youth and Ibn Rushd an old and famous man; the encounter between the two has become somewhat proverbial (see below). Two of Ibn al-'Arabi's maternal uncles (Abu Muslim al-Khawlani and Yahya bin Yughan) were Sufis. His family moved to Seville when he was eight. In Seville, Ibn al-'Arabi received his formal education, which consisted of the study of the Qur'an and its commentary, the traditions of the Prophet, Law (*Shari'ah*), Arabic grammar, and composition. He studied with the great masters of his time and was initiated into the Sufi way in 1184, when he was twenty. Among the subjects taught in the Sufi circles were the metaphysical doctrines of Sufism, cosmology, esoteric exegesis, the science of letters and numbers, and the stages of the Way. In addition, the disciple spent long hours every day engaged in the practices of sufism: invocations, prayers, fasting, vigils, retreat, and meditation.

At thirty, he left the Iberian Peninsula for the first time to visit Tunis. Later he travelled to Fez, then to Alexandria and Cairo, and arrived in Makkah a year before the close of the sixth Islamic century. In Makkah he started to write his enormous compendium of esoteric knowledge, *al-Futuhat al-Makkiyyah*, consisting of 560 chapters. In the year 1204, Ibn al-'Arabi left Makkah and travelled via Baghdad to Mosul, where he spent some time in studying and where he composed a treatise of fifty-three chapters on the esoteric significance of ablution and prayer entitled *al-Tanazzulat al-Mawsiliyyah* (*Revelations at Mosul*). From there he went to Cairo before returning to Makkah in 1207, where he resumed his studies of the Traditions of the Prophet. He stayed in Makkah just over a year and then went northwards to Asia Minor. He arrived in Konya in 1210, where he met his closest disciple, Sadr al-Din al-Qunawi. In 1223 he settled in Damascus, where he died in 1240, leaving behind a circle of disciples whose line continues to this day.

One of the most important terms in Ibn al-'Arabi's cosmology is *wujûd* (existence, being), a term already employed by generations of philosophers before him. In Ibn al-'Arabi's usage, however, this term gained a specific meaning, derived from its basic etymological root, which literally means "finding." It was already a well-accepted doctrine of Islamic tradition that, strictly speaking, this term can only be applied to God, Who alone has true existence; all else that exists does so because its existence has been willed and granted by God; thus, everything other than God derives its existence from God's *amr* (command), and perishes when God no longer supports their existence (*wujûd*).

> The cosmos and the things and beings found in the cosmos are taken as an outward expression of the relations that exist within God and between God and the cosmos. The laws that govern the things and beings of the universe are the same laws that govern the relations among the divine names. To know the cosmos in its full significance, we need to seek out its *roots* and *supports* within God Himself, and these are the divine names, in a broad sense of the term. The things of the cosmos are the names' *properties* or *traces*. Just as the names can be ranked in degrees according to their scope, so also creatures can be ranked in degrees according to their capacity to manifest the properties of the Divine Level. *Ranking in Degrees of Excellence* goes back to roots in God. (Chittick 1998, xix)

To a modern student of cosmology whose understanding is limited to the physical cosmos with no links to the spiritual order, this may seem a difficult and highly subjective approach to cosmology, but those concerned with the *root* of existence cannot but marvel at the penetrating insights into the nature of existence found in Ibn al-'Arabi's cosmological principles. The so-called Laws of Nature often referred to in religion and science discourse are but one set of laws in a hierarchy of principles that govern the cosmos, which is not limited to the physical world studied by modern science but consists of many worlds. This is why the Qur'an refers to God as *Rabb al-'Alamîn*, the Lord of the Worlds. It was mentioned in the introduction that one of the most important aspects of the Qur'anic view of the physical cosmos—and the vast ecological and biological systems operative in it—is that it stands as a sign (*ayah*) and a witness to the One who fashioned it for a purpose and for a fixed duration. Ibn al-'Arabi's cosmology is replete with this understanding. He uses the term *dalil* (indicator) to refer to this aspect of the cosmos. His descriptions of the relationships between God, the cosmos, and humanity lead us to an understanding of the cosmos that is rooted in the central reality of its existence, order, and functioning derived from "the Breath of the All-Merciful" (Chittick 1998, xxvii).

Ibn al-'Arabi's cosmology offers a vast range of intertwined ideas, which integrate various domains of knowledge into a systematic description of existence. There are four basic ways in which he orders the cosmic degrees. The first is in terms of the twenty-eight letters of the Arabic language, beginning with *hamza*, pronounced most deeply in the throat. The other three are temporal (prior and posterior), spatial (higher and lower), and qualitative (more excellent and less excellent) (Chittick 1998, xxix).

We cannot go into further discussion of this important aspect of Islamic cosmology, but let us close this section with a short note on Ibn al-'Arabi's view of causality, for it shows how his cosmology transcends the distinctions and debates that had mired Muslim philosophers for centuries in apparently inextricable discourses. Following Greek thought and terminology, many Muslim philosophers called God the "cause" (*illa*) of the universe and the universe God's "effect" (*ma'lûl*). Perhaps as a result of their excessive preoccupation with these terms, Ibn al-'Arabi sometimes calls these philosophers the "Companions of the Causes" (Chittick 1998, 17). As for himself, he refuses to compromise God's independence at any level:

We do not make Him a cause of anything, because the cause seeks its effect, just as the effect seeks its cause, but the Independent is not qualified by seeking. Hence it is not correct for Him to be a cause.

The cosmos is identical with cause and the effect. I do not say that the Real is its cause, as is said by some of the considerative thinkers, for that is the utmost ignorance of the affairs. Whoever says so does not know *wujûd*, nor who it is that is the Existent. You, O so-and-so, are the effect of your cause, and God is your Creator! So understand! (Chittick 1998, 17)

COSMOLOGICAL DOCTRINES OF THE POST-MONGOL PERIOD

The invasion and destruction of Baghdad by the Mongols ushered in a new era in Islamic thought and science. During this post-Mongol period, all four major schools of Islamic philosophy—the Peripatetic (*mashsha'i*), the Illuminationist (*Ishraqi*), the Gnostic (*Irfani*), and the theological (*Kalam*)—continued to develop in various parts of the Muslim world. Despite the generally held view in the West, which considers Islamic philosophy to have died by the blow served by al-Ghazali's *Tahafut al-falasifah* (*The Incoherence of the Philosophers*) in all parts of the Muslim world except for Islamic Spain, where it is deemed to have survived for a short while due to Ibn Rushd and his rebuttal of al-Ghazali, research during the last few decades has shown the continuation of a rich discourse in many parts of the Muslim world and especially in Iran (Nasr 1997, 20). It is, therefore, not surprising that two of the most important philosophers of the post-Mongol period—Sadr al-Din al-Shirazi (d. 1640) and Haji Mulla Hadi Sabsvari (d. 1873)—were born in Iran. As mentioned in the preceding section, both Suhrawardi and Ibn al-'Arabi brought Peripatetic tradition closer to the Sufi doctrines. Nasir al-Din Tusi on the other hand revived Ibn Sina's school of philosophy, and in so doing he also opened paths for further developments in Peripatetic philosophy that were followed by other philosophers such as his friend and contemporary Najm al-Din Dabiran Katibi (who wrote a major treatise on Peripatetic philosophy entitled *Hikmat al-'ayn (Wisdom of the Fountainhead)*, and, most notably, by Qutb al-Din Shirazi, who was himself a student and colleague of al-Tusi.

The Mosque, the Laboratory, and the Market

Madrassah of Mulla Sadra in Shiraz, Iran. Mulla Sadra's *Transcendent Theosophy Concerning the Four Intellectual Journeys of the Soul* is one of the most important works of Islamic philosophy. © Al-Qalam Publishing.

The period of four centuries between the death of Ibn al-'Arabi in 1240 and the birth of Sadr al-Din al-Shirazi, generally known as Mulla Sadra, is thus not a barren period in Islamic thought but one rich in inner developments that paved the way for Mulla Sadra's grand synthesis of the four major schools of thought. The works of many important philosophers and thinkers of these four centuries have yet to be studied in Western languages. A full picture of the intellectual activity of these four centuries can only become apparent after considerable attention has been paid to the works of such philosophers of these four centuries as Jalal al-Din Dawani, Sadr al-Din Dashtaki, Ghiyath al-Din Mansur Dashtaki, Abd al-Razaaq Kashani, and Dawood Qaysari. What can be said at this stage with confidence is that this period is replete with activity. Mulla Sadra was to benefit from a series of outstanding philosophers preceding him. Notable among them are Mir Damad (d. 1631/32) and Shaykh Baha al-Din Amili (d. 1622). Thus, when the young Mulla Sadra came to Isfahan, he

> entered upon a climate where the intellectual sciences could be pursued alongside the "transmitted" or religious sciences (*al-ulum al-naqliyyah*) . . . when we look back upon the intellectual background of Mulla Sadra, we observe nine centuries of Islamic theology, philosophy and Sufism whch had developed as independent disciplines in the earlier centuries and which gradually approached each other . . . Mulla Sadra was an heir to this vast intellectual treasure. (Nasr 1997, 26–27)

Of the forty-six works of Mulla Sadra, *The Transcendent Theosophy Concerning the Four Intellectual Journeys of the Soul* (*al-Hikmat al-muta'âliyah fi'l-asfâr al-'aqliyyat al-arba'ah*), completed in its first form in 1628, is considered his magnum opus. Its four books discuss all the problems discussed earlier in Islamic theology, philosophy, and Sufism, and since its composition it has drawn the attention of numerous Muslim philosophers. Many commentaries have been written upon it, and it is still an essential work for advanced students of philosophy. Mulla Sadra made a profound distinction between existence (*wujûd*) and quiddity or essence (*mâhiyyah*). Instead of things that exist he made existence the main subject of metaphysics. The distinction between existence and quiddity in Mulla Sadra's system begins with an analysis of how things are ordinarily perceived within the human mind. The concrete objects perceived by the mind have two components:

> One which bestows reality upon the object, which is existence, and the other which determines the object to be what it is, which is its quiddity. Of course in the external world, there is but one reality perceived, but within the mind the two components are clearly distinguishable. In fact, the mind can conceive clearly of a quiddity completely independently of whether it exists or not. Existence is an element added from "outside" to the quiddity. It is not part of the essential character of any quiddity save the Necessary Being, Whose Being is none other than Its quiddity. Existence and quiddity unite and through their union form objects which at the same time exist and are also a particular thing. (Nasr 1997, 103)

Like many Muslim thinkers before him, Mulla Sadra also discussed the question of the eternity of the universe versus a temporal creation in his treatise *Huduth al-'alam*. This treatise discusses creation in time based on Mulla Sadra's doctrine of transsubstantial motion, *al-harkat al-jawhariyyah*, which is one of the basic features of his transcendent theosophy. This principle is used by Mulla Sadra to deal with many questions related to the physical cosmos. He believes that love is the most important and all-pervasive principle of the universe, and therefore, as opposed to Ibn Sina, who denied transsubstantial motion and conceived of becoming as an external process that solely affects the accidents of things, Mulla Sadra "conceives of being as a graded reality which remains one despite its gradation" (Nasr 1997, 91).

The *hikmat* (wisdom) tradition, which achieved great heights through Mulla Sadra, is "structurally a peculiar combination of rational thinking and Gnostic intuition," as Izutsu has described it (Sabzavari tr. 1977, 3). This tradition remains one of the most vibrant schools of philosophy in contemporary Iran. This "spiritualization of philosophy," as it has been aptly called by Izutsu, originated in the metaphysical visions of Ibn al-'Arabi and Suhrawardi, and clearly distinguished a rational and a gnostic component in what was previously merely a logically construed rational discourse. In its logical structure, its philosophical terms, and concepts it takes Ibn Sina's Peripatetic tradition as its point of departure. Its second component, namely a mystical or Gnostic experience, underlies the whole structure of philosophizing. It is this second component of *Hikmat* that makes it a keen analytic process combined with a profound intuitive grasp of reality. Mulla Sadra had realized that mere philosophizing that does not lead to the highest spiritual realization is vain, and it was his belief that there is a reciprocal relationship between mystical experience and logical thinking. This *Hikmat* tradition found a new exponent in the nineteenth century in Mulla Hadi Sabzavari, who was to continue the work of Mulla Sadra. In his *Metaphysics*, Sabzavari points out that existence is self-evident and all defining terms of "existence" are but explanations of the word; they can neither be a "definition" nor a "description" of existence itself.

The foregoing survey shows some of the inherent connections that existed between Islam (the religion) and the science practiced in the Islamic civilization. We have seen that the application of certain contemporary frameworks of studying the relationship between science and religion to Islam and science discourse often produces misconceptions. We have also examined certain key cosmological schemes that emerged in Islamic civilization, and explored how they construed the natural world. This discussion leads to a realization that there is a need to reinvestigate and reexamine certain basic notions about this relationship in a new work based on primary sources.

Islamic scientific tradition was to gradually decline and eventually disappear. When did this decline start? Why? What was done to prevent it? These questions are discussed in the next chapter along with certain aspects of the transmission of science from Islamic civilization to Europe.

Chapter 5

Islam, Transmission, and the Decline of Islamic Science

In this chapter we are broadly concerned with exploring two main questions: (i) What was the role of Islam in the transmission of science to Europe? (ii) Was Islam responsible for the decline of science in Islamic civilization? Both of these questions are intertwined with a host of historical developments in Europe as well as in the Muslim world. Both are complex. Both involve a wide range of individuals and institutions, and both have left deep marks on the subsequent history of Islam and science discourse. In examining these questions, we will limit our discussion to only those aspects that are directly related to the role of Islam in transmission and the decline of science, though certain intertwined issues cannot be left aside in any narrative of this fascinating chapter of human history.

CIVILIZATIONS IN DIALOGUE: THE TRANSMISSION OF ISLAMIC SCIENCE TO EUROPE

The perception most common in the Muslim world about the transmission of science to Europe is the rather untenable claim that the Scientific Revolution of the seventeenth century was a direct result of the transmission of Islamic science to Europe. Conversely, the most common perception in the West about this transmission is that a few Arabic works of Greek origin were translated from Arabic into Latin, but that they were of little use for the rapidly developing science in Europe. Both of these perceptions are erroneous; both are generalizations that have found their way into secondary literature and are repeated ad nauseam; both are exaggerations without historical foundation. Another common misconception arises from comparisons between the translation of Islamic scientific texts into Latin and the translation movement into Arabic. These comparisons are at best superficial, as will become clear from the following account. The two events are of a different order of magnitude, took place under very different conditions, and yielded very different consequences.

The movement that brought Islamic scientific and philosophical thought to the Latin West can be divided into three phases. The first of these three began in the late tenth century as a result of small and individual efforts. The individual most responsible for supporting and spreading this activity was a man born of poor parents in or near the village of Aurillac in south-central France: Gerbert of Aurillac (*ca.* 946–1003). Gerbert, who would rise to the Papacy in 999 as Pope Sylvester II (999–1003), received a thorough education in Latin grammar at the monastery of St. Gerard in Aurillac, where he remained until 967. In that year, Borrell II, the count of Barcelona, visited the monastery and was so impressed by Gerbert that he requested the abbot to allow Gerbert to come to Catalan Spain for further education. In Spain, Gerbert was entrusted to Atto, the bishop of Vich. It was here that Gerbert first came into contact with a mathematics far superior to anything he had learned so far. Fascinated by the Islamic mathematical tradition and Arabic numerals, he quickly learned the use of an abacus, which he later introduced to Latin Europe outside Spain (Lattin 1961, 6).

Gerbert's interest in Islamic scientific manuscripts was sustained over a long period of time, but what is more important for our book is the information we find in his letters about the earliest period of translation activity from Arabic into Latin. For instance, on March 25, 984, he wrote a letter to Seniofred of Barcelona (d. *ca.* 995), the prominent and wealthy archdeacon of the cathedral of Barcelona (also known by his nickname Lupitus or Lubetus (Lobet), in which he asked him to send him *De Astrologia*, his translation of an Arabic text on astrology (Lattin 1961, 69). In another letter, written from Rheims in February or March 984 to Abbot Gerard of Aurillac, he asks for a book on multiplication and division translated by Joseph the Spaniard, a Mozarab who knew Arabic: "Abbot Guarin left with you a little book, *De multiplicatione et divisione numerorum*, written by Joseph the Spaniard, and we both should like a copy of it" (Lattin 1961, 63).

These early tenth-century translations planted the seeds of what became a major intellectual tradition in the next two centuries. This second phase of translation activity (eleventh to the fourteenth centuries), produced a steady flow of Latin translations by a small number of translators, among whom was a Muslim who later converted to Christianity and is known to us as Constantine the African (*fl.* 1065–1085). Not much is known about his life except that he was a merchant who traveled between his home in Tunisia and southern Italy. He became interested in medicine, and after studying it for several years in Tunisia went to Salerno, Italy, carrying with him a large number of Arabic books. Around 1060 he entered the monastery of Monte Cassino, where he stayed until his death in 1087

(McVaugh 1981, 395). He translated a large number of Arabic medical texts into Latin, claiming their authorship for himself. His translations had

a very considerable effect upon twelfth-century Salerno. As the core of the collection entitled *Ars medicine* or *Articella*, which was the foundation of much European medical instruction well into the Renaissance ... it did not merely enlarge the sphere of practical competence of the Salernitan physicians; it had the added effect of stimulating them to try to organize the new material into a wider, philosophical framework. (McVaugh 1981, 394)

While this translation activity was in progress, the reconquest of Spain began in full force. The fall of Toledo in 1080 resulted in the availability of an excellent library to translators. During the previous four and a half centuries (712–1085), *Tulaytulah*, as Toledo was called in Arabic, had become an important center of learning in Islamic Spain. As the language of learning and culture had become Arabic, many non-Muslim residents of Spain too adopted and were fluent in Arabic. Raymond I, the new archbishop of Toledo (1126–1151), patronized the translation movement and gathered around him a small group of translators headed by Archdeacon Dominico Gundislavi (or Gundisalvo), a Spanish philosopher. Gundislavi also wrote a book (*De divisione philosophiae*) that introduced al-Farabi's scheme of classification of knowledge to Europe. Another resident of Toledo, John of Seville (*fl.* 1133–1142), a Mozarab, translated a large number of astrological works into Latin (see Table 5.1). Hugh of Santalla (*fl.* 1145), an astrologer and alchemist, translated al-Biruni's commentary on the astronomy of al-Farghani and also translated works on astrology and divination. Mark of Toledo (*fl.* 1191–1216) translated Galenic texts (Sarton 1931, vol. II, 114). Most translators were from or at least based in southern Europe, as elsewhere there was little access to manuscripts or major interest in translation activity.

Table 5.1 gives details of major translations done during this second phase (eleventh to the fourteenth centuries). These three centuries saw considerable expansion of the links between the Muslim world and a new Europe emerging from the ruins of the Roman Empire through a complex process of intermingling of populations, Viking settlements, and unprecedented economic growth. Along with the emergence of stable monarchies and population explosion, there arose a chain of new schools throughout Western Europe with far broader aims than the previous monastery schools had. These new schools were centered on the interests of the Master who directed them (just like schools in Islamic civilization, which attracted students to a particular teacher whose name was synonymous with that of the school). And just like their counterpart in the Muslim world, these European schools were not geographically fixed; they went where their master-teacher went. The number of students and teachers in these new

Table 5.1
Some Arabic scientific and philosophical works translated into Latin between the eleventh and the thirteenth centuries

Author	Arabic Work	Latin/English Title	Translator
al-Khwarizmi	Astronomical tables	*Ezich Elkauresmi per Athelardum bathoniensem ex aribico sumptus*	Adelard of Bath (*fl.* 1116–1142)
Abu Ma'shar	*Kitab al-madkhal al-Saghir (Shorter Introduction to Astonomy)*	*Ysagoga minor Iapharis matematici in astronomicam per Adhelardum bathoniensem ex Arabic sumpta*	Adelard of Bath
Ibn Sina	*Risala (maqala) fi-l-nafs*	*De anima*	Gundisalvo with Ibn Dawud
al-Ghazali	*Maqasid al-falasifa*	*The Aims of the Philosophers*	Gundisalvo with Johannes
al-Farghani	*'Ilm al-nujum*	*De Scientia astorum*	John of Seville (*fl.* 1133–1142)
Abu Ma'shar	*Kitab al-Madkhal al-kabir ila 'ilm ahkam al-nujum*	*Great Introduction to the Science of Astrology*	John of Seville
Qusta bin Luqa	*Kitab al-Fasl bayn al-ruh wa-l-nafs*	*De differentia spiritus et anime*	John of Seville
al-Khwarizmi	*Algebra*	*Algebra*	Robert of Chester (*fl.* 1140–1150)
al-Kindi	*al-Kitab al-astarlab*	*De iudiciis astrorum*	Robert of Chester
al-Battani	*Zij al-sabi*	*De motu stellarum*	Plato of Tivoli
Ibn Sina	*Al Qanun fi'l-tibb*	*Canon of Medicine*	Gerard of Cremona (ca. 1114–1187)

schools increased rapidly and some of them became large enough to require organization and administration, leading to the emergence of the universities that would subsequently become home to intense scientific activity.

Earlier translations of this second phase were done without a scheme or definite plan, but as translated texts began to circulate specific needs arose for the translation of works referenced in earlier translations. Among the greatest translators was Gerard of Cremona (d. 1187), an Italian who came to Spain in the late 1130s or early 1140s in search of Ptolemy's *Almagest*. He found a copy in Toledo. Once in Toledo, "seeing the abundance of books in Arabic on every subject ... he learned the Arabic language, in order to be able to translate" (Lemay 1981, 174). Over the next three decades, Gerard was to produce over eighty translations of scientific and philosophical texts from Arabic, no doubt with the help of a team of assistants. These translations are not of high quality, but their importance lies in the introduction of a vast corpus of Islamic scientific and philosophical texts to Latin scholars and in their subsequent impact on the history of science as well as on the discourse between Christianity and science.

It is important to have a closer look at what Gerard translated, for it sheds light on the interests of Latin scholars of the time. It also provides us important clues to understand the nature of subsequent discourse on the relationship between science and Islam on the one hand and science and Christianity on the other. Eighty-two works listed in the incomplete bibliography of his works prepared by his companions in Toledo can be divided into six categories: (i) logic—three works; (ii) geometry, mathematics, optics, weights, dynamics—seventeen works; (iii) astronomy and astrology—twelve works; (vi) philosophy—eleven works; (v) medicine—twenty-four works; (vi) alchemy—three works; (vii) geomancy and divination—four works (Lemay 1981, 173–92). Among the originally Greek works translated from Arabic are the *Posterior Analytics* of Aristotle, Ptolemy's *Almagest*, and Euclid's *Elements*. Works by Muslims translated by Gerard and his companions include al-Khwarizmi's *Algebra*, al-Farabi's *Short Commentary on Aristotle's Prior Analytics*, Banu Musa's *Geometria*, al-Kindi's *De aspectibus* (optics), Thabit ibn Qurra's *liber Qarastonis*, twelve works on astronomy and astrology including al-Farghani's *Liber Alfagani in quibusdam collectis scientie astrorum ett radicum motuum planetarum et est 30 differntiarum*, Jabir ibn Aflah's *De astronomia libri IX*, and several works by al-Razi, including his *Liber Almansorius*, the shorter of al-Razi's great medical works.

The future European discourse on Islamic scientific tradition was shaped to a certain extent by what was translated at this time. It is interesting to note that these translations were done at a time when many important contributions of the Islamic scientific and philosophical traditions had yet to appear. It is also interesting to note that Gerard and his companions paid no attention to a man who would become the most important Muslim philosopher for Europe and who lived close to the time of their translation activity: Ibn Rushd (d. 1198). His works were later translated in the thirteenth century, shortly after his death. Gerard is, nevertheless, one of the most important transmitters of knowledge from one civilization to another and is at times compared to Hunayn ibn Ishaq (d. 873), the Nestorian Christian who translated some 129 works from Greek and Syriac into Arabic and whom we have mentioned in Chapter 2. Another important aspect of this second phase of translations from Arabic into Latin is the absence of any substantial link between the translators of this period and the Muslim scholars and scientists who lived in the eastern parts of the Muslim world. Most of this translation activity was based on what was available in Spain.

The last page of Ibn Sina's magnus opus in medicine, *Al-Qanun fi'l tibb* (*The Canon of Medicine*) which was translated into Latin by Gerard of Cremona (*ca.* 1114–1187) and which remained the standard textbook of medicine in Europe until the sixteenth century. © Al-Qalam Publishing.

During the thirteenth century, William of Moerbeke (*fl.* 1260–1286) rendered the entire Aristotelian corpus into Latin, along with many works of Aristotle's Muslim commentators. He also revised older translations and translated a number of neo-Platonic works. The most important translation for the science and religion discourse in Europe, however, was yet to come: the translations of Ibn Rushd by Michael Scot, a Scotsman who inaugurated Averroism into the European tradition. Among other translations of note from this century are Alfred Sareshel's translation of the alchemical part of Ibn Sina's *al-Shifa* and Michael Scot's translation of al-Bitruji's work on astronomy in 1217.

In astronomy, the most important work of this century was *The Alfonsine Tables*, drawn up at Toledo around 1272 by order of the king of Castile and León, Alfonso X (d. 1284). Alfonso was the son of Ferdinand III (d. *ca.* 1252), the conqueror of some of the most important cities of Muslim Spain including Cordoba, Murcia, and Seville. Alfonso's entourage included many scholars, both Christian and Jewish, who knew Arabic. He patronized translations from Arabic into Castilian, making it a new vehicle for scientific communication. These new tables extended the scope of translations from Arabic. Their impact on the Latin astronomical tradition should be seen in the context of an already existing Islamic influence on Latin astronomy due to translations of the astronomical tables of al-Khwarizmi and other Muslim scientists. *The Alfonsine Tables* took full advantage of the earlier translations, in particular building upon the *Toledian Tables*, a compendium of Islamic astronomical tables compiled by Said al-Andalusi and his circle and translated into Latin in the twelfth century.

The Alfonsine Tables divided the year into 365 days, 5 hours, 49 minutes, and 16 seconds. Once translated into Latin (in 1320s Paris) they remained the most popular astronomical tables in Europe until late in the sixteenth century, when they were replaced by Erasmus Reinhold's Prutenic Tables based on Copernicus's *De revolutionibus orbium coelestium* (Chabás and Goldstein 2003, 243–247).

One of the most important figures in Toledo translations at the time of Alfonso's reign was Gonzalo García Gudiel, who established his own scriptorium around 1273 (Chabas and Goldstein 2003, 226). Table 5.2 lists a sampling of Arabic works translated into Castilian under the patronage of Alfonso.

This second phase of the translation movement was contemporaneous to the founding of several new universities in Europe. These include the Universities of Bologna (1150), Paris (1200), and Oxford (1220). (These dates are approximate, as the process of achieving university status involved several steps, and dating depends on which step is considered most significant.) What was taught in these new universities changed over time, but an interesting feature of these early universities was the uniformity of curriculum. There were minor differences in emphasis but almost all universities taught the same subjects from the same texts. This may simply have been the result of the sheer paucity of texts available, but this common curriculum produced a phenomenal result: medieval Europe acquired a universal set of Greek and Arabic texts as well as a common set of problems that facilitated a high degree of student and teacher mobility across the continent. Thus, teachers earned their right of teaching anywhere and moved between different universities, all of which used Latin as their language of instruction. This demonstrates an important parallel between medieval Europe and the Muslim world, where

The Cordoba Mosque is an important remaining monument of Islamic Spain (al-Andalus). It became the center of transmission of Islamic scientific tradition to Europe during the eleventh to fourteenth centuries. © Al-Qalam Publishing.

Table 5.2
Some of the Arabic scientific and philosophical works translated into Castilian

Date	Arabic Title	Castilian/English	Translator
ca. 1263	*Kitab al-mi 'raj*	*The Book of Muhammad's Ladder*	Abraham of Toledo
1254	An unknown work by Abu l-Hasan 'Ali ibn Abi l-Rijal	*Libro complido en los iudizios de las estrellas*	Judah ben Moses ha-Cohen
1256	The star catalogue for 964 by Abu l-Husain 'Abd al-Rahman ibn 'Umar al-Sufi	*Libro de las estrellas de la ochaua espera*	Judah ben Moses ha-Cohen with the help of Guillen Arremaon Daspa
1259	A treatise on the use of a celestial globe written in the ninth century by Qusta ibn Luqa	*Libro de la alcora*, also called *Libro de la faycon dell espera et de sus figuraas et de sus uebras*	Judah ben Moses ha-Cohen, in collaboration with Johan Daspa; a revised version completed in 1277
1259?	The treatise on the astrolabe by Abu l-Qasim ibn al-Samh (d. 1035), a disciple of Maslama al-Majriti	*Libro del astolabio redondo*	Book I was compiled by Isaac ben Sid
1258?	A treatise 'Ali ibn Khalaf on the construction and use of a universal astronomical instrument using meridian projections	*Libro de la lamina universal*	Compiled by Isaac ben Sid
1255–1256	Treatise by Azarquiel on the construction and use of a *saphea* of the *zarqaliyya* type	*Libro de la acafeha*	First translated by Fernando de Toledo, but found unsatisfactory by the King; a new version made in 1277 by Bernardo "el arauigo" and Abraham of Toledo
1259	An astrological treatise from eighth-century al-Andulus	*Libro de las cruzes*	Judah ben Moses ha-Cohen with the help of Johan Daspa
After 1270	Ibn al-Haytham's *Kitab fi hay'at al 'lam*	*On the Configuration of the World*	Abraham of Toledo

111

Arabic was the universal language of scholarship and where students and teachers easily moved across a vast geographical expanse. Translations from Arabic provided an important source of texts used in these universities.

The third phase of translation activity can be classified as roughly belonging to the period between the sixteenth and seventeenth centuries. It differs from the first two phases in many important respects. Its scope is much wider, both geographically as well as in terms of material. Instead of courts, this phase of translation activity was increasingly based in the newly founded universities. Translators were now able to travel to Muslim lands other than Spain. The translations from this phase are much more refined and critical. But the most important aspect of this phase of the translation activity was that it gave birth to a distinct enterprise that would cast a deep shadow on the West's relationship and understanding of Islam and Muslims: Orientalism. Because of its importance, this third phase of translation activity deserves more detailed exploration and a context.

EUROPEAN CONTEXT

In order to holistically understand the three phases of European translation activity it is important to keep in mind that the second phase of translation activity took place in the backdrop of the loss of al-Andalus by Muslims through its reconquest by the Latin West and, more important, of the Crusades. Both events contributed to the emergence of a very unfavorable image of Islam and Muslims in the European mind. Islam, for most Europeans of that time, was a dangerous, hostile, and even pagan cult. This image was built through another translation movement that began somewhat prior to the translation of scientific texts and focused instead on Islamic texts, transmitting to the learned Latin circles of the Middle Ages (*ca.* 800–1400) an account of Islam's message and the life of its Prophet. The transmitters of this information lived close to Muslims and had access to Islam's two primary sources, the Qur'an and *Hadith*.

The image of Islam prevalent in the medieval West emerged on the basis of the works of these writers. In many writings of this period, the Prophet Muhammad appears as an idol worshipped by Saracens; in others he is depicted as a magician; in still others he is a possessed man. In the epics of the Crusades, the Prophet of Islam appears as a heathen god (Trude 1993, 382). These popular texts had an audience not trained in theological intricacies, and their authors and minstrels used imaginative powers to exaggerate and hyperbole for their readership; this resulted in such gross misrepresentations as the infamous *Mary Magdalene* from the Digby cycle. An amazing characteristic of the portrayal of the Prophet Muhammad in the popular texts of the Middle Ages is the consistent negative image in languages as far removed from each other as Old Icelandic, German, and English. This is because the original material for the popular texts was found in common Latin or Arabic sources. In popular English literature, for instance, Prophet Muhammad was represented as a renegade cardinal in *Piers Plowman* of William Langland (1362); in John Lydgate's *The Fall of the Princes* (1438) he appears as a heretic and false prophet in the story *Machomet the false Prophete*.

As the intellectual milieu of Europe at the time of the translation of Islamic scientific texts was thus influenced by this parallel translation movement, the Crusades shaped the social outlook of Europe of the time. They began when Emperor Alexius I, having suffered a defeat by the Abbasid army, appealed to Pope Urban II for help. He requested that the Pope undertake a pilgrimage to liberate Jerusalem from the Muslims. In 1095, Urban II gave the call for the holy war and the Papal battle cry *Deus vult!* ("God wills it!") resounded throughout Europe. Knights, merchants, and ordinary soldiers marched toward Jerusalem. In 1099, Godfrey of Bouillon captured the city; his men massacred almost the entire Muslim and Jewish population of the holy city. The incident shocked the Muslim world but did not pose a serious threat to the vast Abbasid Empire. In 1187 Salah al-Din (d. 1193) recaptured Jerusalem. The Crusades, however, continued until the end of the thirteenth century, when they degenerated into intra-Christian wars.

The Crusades are more important for their effect on the perception of Islam and Muslims in the West than their military victories. For almost two centuries, mesmerized by the Papal battle cry, allured by promises of rich booty and heavenly rewards, and driven by the attraction of the holy city, thousands of men marched through various towns and cities on their way to Jerusalem. This had an invigorating effect on the popular mind, which saw Muslims as the barbaric infidel. It was also during this time that the average European learned the most horrible details about Islam and its Prophet in a climate already charged with hatred and mistrust. It was in this environment that the earliest polemics against Islam were written in Europe, based in part on information received from al-Andalus. The purpose of these writings was to justify the "holy war." Muslims appear in these writings as polytheistic, polygamous, promiscuous, worshippers of Muhammad, and wine-drinkers. These images passed from generation to generation and their remnants are still reflected in certain perceptions about Islam and Muslims held in the West.

The European Renaissance attempted to rebuild a civilization based on its antiquity. This incessant return to the past is apparent everywhere—in art, in the sciences, in poetry. At the same time, European intellectuals, writers, scientists, and artists of this period developed an intense hatred for Islam, its Prophet, and his family. Dante Alighieri (1265–1321), for instance, placed Ibn Sina and Ibn Rushd in Limbo, in the First Circle of

Hell, with the greatest non-Christian thinkers—Electra, Aeneas, Caesar, Aristotle, Plato, Orpheus, and Cicero—where they must live without hope of seeing God, in perpetual desire, though not in torment (Dante 1971, Canto IV: 142–144). But he placed the Prophet and his son-in-law, Ali, among a group of "sowers of scandal and schism," whose mutilated and bloody bodies, ripped open and entrails spilling out, bemoan their painful lot: "See how Mohomet is deformed and torn!/In front of me, and weeping, Ali walks,/his face cleft from his chin up to the crown" (Dante 1971, Canto XXVIII: 31–33).

The first Latin paraphrase of the Qur'an, made by Robert of Ketton at the behest of Peter the Venerable, Abbot of Cluny, was completed in 1143. (It still exists with the autograph of the translator in the Bibliothèque de l'Arsenal in Paris.) An Italian version was published by Andrea Arrivabene in 1547, and "though its author claims that it is made directly from the Arabic, it is clearly a translation or paraphrase of Robert of Ketton's text, published by Bibliander. Arrivabene's version was used for the first German translation made by Solomon Schweigger, which in turn formed the basis of the first Dutch translation, made anonymously and issued in 1641" (Pearson 1986, 431). Most of the subsequent translations of the Qur'an in various European languages were derivative products of these works (which were themselves not accurate in the first place), and it was not until the second quarter of the twentieth century that Muslims produced their own translations of the Qur'an in European languages. With this background in mind, let us now examine the third phase of translation.

THE THIRD WAVE OF TRANSLATIONS

The sixteenth and seventeenth centuries have been described as the "golden age of Arabic studies in Europe" (Feingold 1996, 441). During these two centuries, a third wave of translations emerged along with a more serious interest in things Islamic. This was institutionalized in the form of several professorships of Arabic founded in European universities. During these two centuries, "scores of scholars made their way East in search of instruction in the language or for Arabic manuscripts—thousands of which made their way to Europe—and various publishers as well as individual scholars acquired Arabic type in anticipation of a significant publication enterprise" (Feingold 1996, 441). The translation activity of this period produced annotated translations of Arabic texts, often along with the original Arabic. This marked interest in primary sources, while indicative of a mental attitude formed by reformation and humanism (as noted by Feingold), was also entangled in the intellectual and theological debates that proliferated in Europe during the sixteenth and the seventeenth centuries. In any case, through patronage, internal politics of the European academic community, and necessity, the study of

Arabic did become indispensable to the late Renaissance humanists, who applied it to gain access to their cherished classical texts preserved and enriched by the Muslim scholars. Although this phase of the transmission of knowledge from the Islamic civilization to Europe remains least studied, it clearly shows great interest—even zeal—in learning Arabic and in acquiring works of Islamic tradition as late as the seventeenth century. The teaching of Arabic became an integral part of the academic curricula through the establishment of chairs, research programs, and several ambitious projects.

Two new factors contributed to the renewed interest in the study of Arabic during the sixteenth and the seventeenth centuries: hostile interaction between the Ottomans and Europe and the growth of interest among Western Europeans in establishing contacts with the Eastern Churches, which used Arabic, Greek, Syriac, Turkish, and Coptic as their liturgical and/or vernacular languages (Toomer 1996, 7–15). The study of Arabic was thus pursued for two motives: the acquisition of scientific and philosophical texts, and for Christian missionary and apologetic activities. France had the distinction of being the first European country to establish formal relations with the Ottoman Empire and to institute formal instruction in Arabic, both under King François I. Many influential scholars encouraged the study of Arabic in public lectures. The chairs of geometry and astronomy established at Oxford in 1619 required knowledge of Islamic scientific tradition as an essential qualification. A succession of European scholars produced a sustained flow of translation activity during these two centuries. More important, however, is the fact that these scholars were now themselves going to the Muslim world to perfect their Arabic and collect new manuscripts and first-hand information about Islam and Muslims. Thus Bedwell, Selden, Bainbridge, Pococke, and Greaves—among others—considerably increased European knowledge about Islam and its intellectual tradition, including the natural sciences.

The nature of European interest as well as European perception of Islam and Muslims was, however, going to change drastically in the seventeenth century. This was primarily due to the fact that European scientific and philosophical traditions were now able to surpass the received material from Islam. This radical change in attitude was, however, broader in nature and not confined to scientific learning. Islam and Muslims in general were going to be relegated to second-class status—a position that remains entrenched in Western scholarship to this day. Numerous factors contributed to this change. Arabic texts were no more marvels of wisdom and knowledge, as John Greaves's 1646 complaint to Pococke indicates:

To speak the truth, those maps, which shall be made out of Abulfeda, will not be so exact, as I did expect; as I have found by comparing some of them with our modern and best charts. In his description of the Red sea, which was not far from him, he is most grossely mistaken; what may we think of places remoter? However, there may be good use made of the book for Arabian writers. (Feingold 1996, 448)

A similar sentiment can be discerned from a letter of Robert Huntington, written from Aleppo on April 1, 1671, to John Locke: "The Country is miserably decay'd, and hath lost the Reputation of its Name, and mighty stock of Credit it once had for Eastern Wisedome and learning; It hath followed the Motion of the Sun and is Universally gone Westward" (Toomer 1996, v).

By the end of the seventeenth century, new publications had started to appear on the basis of previously translated Latin texts, which provided foundations for the emergence of Orientalism. As European science surpassed Islamic scientific tradition, the propagandists of the new science began to single out Arabs as harbingers of scholasticism, mere imitators of the Greeks, whose learning was derivative and irrelevant. "The sciences which we possess come for the most part from the Greeks," wrote Francis Bacon (1561–1626) in *Novum Organum*, "for what has been added by Roman, Arabic, or later writers is not much nor of much importance; and whatever it is, it is built on the foundations of Greek discoveries" (Bacon 1905, 275). Bacon's verdict was to become entrenched in subsequent centuries. For him, "only three revolutions and periods of learning can properly be reckoned; one among the Greeks, the second among the Romans, and the last among us, that is to say, the nations of Western Europe, and to each of these hardly two centuries can be assigned. The intervening ages of the world, in respect of any rich or flourishing growth of sciences, were unprosperous. For neither the Arabians, nor the Schoolmen need be mentioned; who in the intermediate times rather crushed the sciences with a multitude of treatises, than increased their weight" (Bacon 1905, 279). Bacon's verdict has been repeated time and again and continues to be the main thesis of most mainstream literature on Islamic scientific tradition. "George Starkey criticized all of the Arabic writers because of their reliance on Galen and opined that 'Avicenna was useless in the light of practical experience'" (Greaves 1969, 90).

By the turn of the eighteenth century, not only the scientific learning of Muslims but Muslims themselves were the subject of judgments: "It is certain that the Arabs were not a learned People when they over-spread Asia," wrote William Watton (1666–1727), "so that when afterwards they translated the *Grecian* Learning into their own Language, they had very little of their own, which was not taken from those Fountains" (Watton 1694, 140).

This tradition of censure first appeared among the humanists and was built upon by the historians of philosophy in the seventeenth century.

Frontispiece of Johannes Hevelius's *Selenographia* (1647), showing Ibn al-Haytham (d.1040) on the left and Galileo (1564–1642) on the right as two representatives of science. © History of Science Collections, University of Oklahoma Libraries.

Leonhart Fuchs demanded the liberation of medicine "from the Arabic dung dressed with the honey of Latinity"; then he went on to state,

> I declare my implacable hatred for the Saracens and as long as I live shall never cease to fight them. For who can tolerate a past and its ravings among mankind any longer—except those who wish for the Christian world to perish altogether. Let us therefore return to the sources and draw from them the pure and unadulterated water of medical knowledge. (Feingold 1996, 442)

In sum, initially material translated from the Islamic scientific and philosophical traditions was deemed necessary for the development of science in Medieval Europe. Later a change took place in European attitude toward Islamic scientific tradition. The colonization of the Muslim world also contributed toward European attitudes toward Islam and its tradition of learning, including the sciences. This colonizer–colonized relationship also affected European attitudes toward Islam and Muslims. In the meanwhile, the Muslim world itself was going through basic internal changes that drastically weakened the Islamic scientific tradition and finally choked it altogether. The next section explores this process of withering.

ISLAM AND THE DECLINE OF SCIENCE

Since this book is mainly concerned with the relationship between Islam and science, we will limit discussion on the historical demise of Islamic scientific tradition to the question mentioned at the beginning of the chapter: was Islam responsible for the decline of science in Islamic civilization? To explore this question, it is necessary to examine certain related questions: What is meant by the decline of Islamic scientific tradition? When did it take place? Where? Did it take place all across the Muslim world at the same time or did it happen in stages? Did all branches of science suffer the same fate at the same time or was it a stepwise process?

These questions have of course been asked by historians of science. The answers vary depending on the perspective taken by the author as well as his or her personal inclinations, ideological commitments, and general attitude toward Islam. The perspective and personal preferences can sometimes even alter the question. Thus, instead of looking into the process of decline of Islamic scientific tradition, some scholars have tended to overshadow this question with another: Why did Muslim scientists not produce a scientific revolution like that which took place in Europe? This formulation radically changes the inquiry, for now the enterprise of science in Islam is being examined against a preset benchmark belonging to another civilization. Even in formulations where this benchmark is

Imam Square in Isfahan (Iran) is a fine example of the vigor of Islamic scientific tradition during the Safavi period. Masjid Imam displays many examples of a high degree of scientific and technological knowledge. Designed by Shaykh Baha'I, the Mosque has one specific spot inside the large prayer area from where sound resonates throughout the Mosque. © Al-Qalam Publishing.

not so obvious, it often remains just below the surface. Perhaps this is unavoidable or even natural, because Islamic scientific tradition immediately preceded the emergence of the Scientific Revolution in Europe.

To begin with, we must ask: what is actually meant by the decline of a scientific tradition? Obviously it is not something like the death of an individual, which happens at eleven in the morning on the first day of the fifth month of a certain year. What we should be looking for is, therefore, a period of time during which the enterprise of science over a large area declined. It also means that this period of time cannot be identical for all regions of the Muslim world. After all, we are dealing with an enterprise that had different local patrons and institutions in different regions of the Muslim world. It also follows that we must ask: If this decay and decline was a slow process over a certain period of time, were there any attempts to cure the malady? If yes, what were they? Who made them? What was the role of religion in this process? Given that science—any science—does not exist in isolation; it follows, then, that we must inquire about intellectual, social, economic, and political conditions of the Muslim world during the period of decline and attempt to see certain relationships between these broad conditions and science. We cannot do justice to these questions without the discovery, annotation, and publication of a large number of manuscripts pertaining to the social, economic, and political situation during the period of decline. Nor can we begin to formulate any theory of decline in the absence of a rigorously documented history of the Islamic scientific tradition.

Answers have been provided, often with commanding authority, despite the lack of fully documented source material. These answers vary, but most have a common denominator in that they trace the main reason for both the lack of a Scientific Revolution in Islamic civilization as well as for the demise of its own scientific enterprise to Islam itself. This perspective originated in Europe and has now become the mainstay of Western scholarship on Islamic scientific tradition. The general acceptance granted this answer makes obligatory new work devoted to carefully examining the evidence provided for this answer. This is an unpleasant duty, for it burdens the writer with the task of asking the jury to reopen the case after a verdict has already been pronounced.

WHEN DID THE DECLINE TAKE PLACE?

The existing literature on dating the decline of Islamic scientific tradition mention dates that differ not by years or decades but by centuries. When Edward Sachau translated al-Biruni's monumental *Chronology of Ancient Nations* in 1879, he marked the tenth century as the "the turning point in the history of the spirit of Islam," and made al-Ash'ari and al-Ghazali the culprits: "But for Al Ash'ari and Al Ghazali the Arabs might have been a nation of Galileos, Keplers, and Newtons" (al-Biruni tr. 1879, x). In the 1930s and 1940s, when George Sarton wrote his monumental work, *An Introduction to the History of Science*, he set the eleventh century as the end of the vigor of the Islamic scientific tradition, with the twelfth century and to a lesser extent the thirteenth century as the centuries of transition of that vigor to Europe (Sarton 1931, vol. 2, 131–48). But within two decades of the publication of his work, the discovery of new texts pushed this boundary further, and eventually the entire question of dating the decline had to be recast.

What we know now allows us to say with confidence that the work of astronomers and mathematicians such as Athir al-Din al-Abhari (d. *ca.* 1240), Mu'ayyad al-Din al-Urdi (d. 1266), Nasir al-Din al-Tusi (d. 1274), Qutb al-Din al-Shirazi (d. 1311), and Ibn al-Shatir (d. 1375) cannot be discounted as isolated examples of individual scientists pursuing first-rate science in the thirteenth and the fourteenth centuries. Similarly, the work of al-Jazari (d. *ca.* 1205) in mechanics, of Ibn al-Nafis (d. 1288) in medicine, and of Ghiyath al-Din al-Kashi (d. 1429) in astronomy are testimonies to a living and vibrant scientific tradition as late as the fifteenth century. In addition, we have new evidence from studies on Islamic scientific instrumentation which provides an altogether different methodology for understanding the question of decline, as David King has recently shown by his study of astronomical instruments, which were being made of the highest caliber in Iran in the first decades of the eighteenth century (King 1999, xiii). These examples can be multiplied to include many other scientists and instruments.

Interestingly, one view on decline implicitly suggests that there was, in fact, no such thing as decline of the Islamic scientific tradition, because no such tradition ever existed. All that Islamic civilization did, the advocates of this view argue, was to "host" the Greek tradition it received through the translation movement for three centuries, during which science failed to take roots in the Muslim world because of fierce opposition from religious scholars. Greek science was then transferred to Europe, where it found its natural home and blossomed to become modern science. We have already dealt with this view in Chapters 1 and 2. A variant of this extreme view is the "marginality thesis"—cogently formulated by A. I. Sabra (1987)—which limits the practice of natural sciences in Islamic civilization to a small group of scientists who had no social, emotional, spiritual, or cultural ties with Islamic polity and who practiced their science in isolation. While Sabra has attempted to show "the falsity of the marginality thesis ... by offering a description of an alternative picture—one which shows the connections with cultural factors and forces, thereby explaining (or proposing to explain) not only the external career of science and philosophy in Islam, but at least some of their inherent characteristics, possibilities and limitations" (Sabra 1994, 230), his refutation remains limited to a "few general remarks." But a more serious problem with this refutation is its acceptance of the "two-track thesis," which is itself the cornerstone of the marginality hypothesis Sabra tries to refute. The two-track thesis views the Islamic scientific tradition in opposition to—or at least in competition with—what it calls the Islamic religious sciences. The mainstay of these arguments is phrases such as "sciences of the ancients," which sometimes occur in certain Islamic texts.

What came into the main currents of Islamic thought from outside was received into the Islamic intellectual tradition with critical appraisal and sorting; all living traditions do this. The problem arises when one construes the arrival of the new sciences as if it were arrival *en masse*, like some kind of single body, which met severe opposition from the upholders of traditional Islam.

The Greek scientific tradition was translated into Arabic over a period of three centuries. This process involved numerous influential persons who had diverse temperaments, racial and intellectual backgrounds, and reasons for their involvement in this task. The translation movement did not solely bring Greek science but also brought Indian and Persian scientific data and theories into Arabic. Whatever came into Islamic tradition went through several levels of transformation, ranging from instant linguistic to more fundamental and substantial transformations of content and its metaphysical underpinnings. Thus, to construe the Islam and science relationship as a case of an imaginary Islamic orthodoxy versus a small group of scientists is to distort historical data.

Returning to the question of dating decline, it seems best to look at individual branches of science and different regions of the Muslim world, for even a cursory glance at the scientific data indicates that different branches of science had different high periods and declines. Astronomy flourished at the beginning of science in the Islamic tradition, went through a static period, and then burst once more onto the forefront. Alchemy had initial energy and then remained unchanged for several centuries. Mathematics developed steadily throughout the eight hundred years, as did geography and geology. Most of all, what needs to be recognized is that we simply do not as yet have enough source material to pass any conclusive judgment.

We cannot pronounce a general death sentence to all branches of science in all regions of the Muslim world at a specific date. The need is to carefully study available data (with the understanding that we do not possess all manuscripts and instruments) pertaining to different branches of natural science in different regions of the Muslim world, look at the evidence from within each branch of science to determine its high and low points of productivity, and then categorize a time period during which its study

Ali Qapu Palace in Imam Square, Isfahan, Iran, was the official residence of the Safavid king, Shah Abbas. The entire old city was visible from the roof of the Palace. The large pool on the roof-terrace of the Palace was fed through clay pipes from neighboring mountains. © Al-Qalam Publishing.

declined. This is a task for historians of science who have adequate linguistic and scientific expertise. Even then this judgment will be provisional until a substantial number of new manuscripts have been studied, for, as King has pointed out, so far we only know of about 1,000 Muslim scientists

Lutfullah Mosque, Isfahan, Iran, was the private mosque of the Safavid rulers. It was connected with the Palace through an underground tunnel. © Al-Qalam Publishing.

who worked between eighth and the eighteenth centuries; there are thousands more about whom we have no information or of whom we merely know the names and their works' titles. There are over 200,000 manuscripts in Iran alone, of which about three-quarters are as yet uncatalogued. "In 1994," King writes, "during my research on the first world-map, the index to a 21 volume catalogue of over 8,000 manuscripts in the public library of Ayatallah al-Uzma Ma'rashi Najafi in Qum landed on my desk. There are over 400 titles relating to mathematics and astronomy, including some of the works hitherto thought to be lost" (King 1999, 4, n. 4 and 5). King's book alone cites 9,002 instruments, over 80 manuscripts, and 38 pages of bibliography.

THE "WHY" QUESTION

The "why" question is much harder to answer. Existing literature about the causes of decline of science in Islamic civilization tells us that it was due to (i) opposition by "Islamic orthodoxy"; (ii) a book written by Abu Hamid al-Ghazali; (iii) the Mongol invasion of Baghdad in 1258; (iv) the lack of institutional support for science; (v) the disappearance of patrons;

or (vi) some inherent flaw in Islam itself. This puzzling array of causes—though the list is by no means complete—has been cited in respectable academic publications in a decisive, authoritative manner, with citations and references to support these claims. Yet none of this advances our understanding of this complex question; all we have is opinions of various authors supported with selective evidence often removed from its context.

Those who wish to examine the question of decline in relation to the Scientific Revolution further complicate the matter by reading future developments back into history. Given this situation, the most important task for a new work of the present kind is to attempt to clear existing confusion and point to the possible areas of future research that can provide a more satisfactory answer to the question of decline of science in Islamic civilization.

PREVALENT VIEWS ON THE DECLINE OF SCIENCE IN ISLAMIC CIVILIZATION

The architectonics of much of the Western academic writings dictate a framework in which most new works are based on previous works. When new ideas arise, they arise due to a quantum leap in someone's understanding of the subject, or as refutations of existing ideas. In the case of Islam in general, and Islam and science discourse in particular, such quantum leaps have been almost nonexistent. Thus, what we have is an intolerable repetition, always going back to Goldziher's 1916 formulation. Here is an example.

"During the thirteenth and fourteenth centuries, Islamic science went into decline; by the fifteenth century, little was left. How did this come about?" asks David Lindberg in his 1992 book, *The Beginnings of Western Science*. While he admits "not enough research has been done to permit us to trace these developments with confidence ... several causal factors can be identified" (Lindberg 1992, 180). The first factor is none other than what Goldziher identified in 1916: "conservative religious forces." The second is the "debilitating warfare, economic failure, and the resulting loss of patronage" without which "the sciences were unable to sustain themselves" and the third is, once again, a repeat of Goldziher's basic thesis: "In assessing this collapse, we must remember that at an advanced level the foreign sciences had never found a stable institutional home in Islam, that they continued to be viewed with suspicion in conservative religious quarters, and that their utility (especially as advanced disciplines) may not have seemed overpowering" (Lindberg 1992, 181).

Cohen in his *The Scientific Revolution: A Historical Inquiry* first tells us, "the upshot of all this is that, in 1300, the world of Islam looked quite different from three centuries previously" (Cohen 1994, 408). His proofs come from an essay by J. J. Saunders, which in turns leads us back to Goldziher:

The free, tolerant, inquiring and 'open' society of Omayyad, Abbasid and Fatimid days had given place, under the impact of the devastating barbarian invasions and economic decline, to a narrow, rigid and 'closed' society in which the progress of secular knowledge was slowly stifled. (Saunders 1963, 701–20)

Saunders adds further that "no more borrowings took place from the world outside Islam; Hellenistic philosophy now came to be seen as a danger to the faith. All this was codified in writing by al-Ghazali, and Ibn Rushd's defence of the *awâil* sciences, one century later, failed to carry conviction" (Saunders 1963, 408). Even ignoring the unmistakable Goldziherism (*awâil* sciences versus Islam as well as al-Ghazali against science), there are conceptual problems as well. For example, Saunders's prognosis above would have Islamic civilization keep on borrowing from the world outside Islam even after the three hundred years old translation movement had exhausted available outside sources!

There is a long history of such texts, almost all of which were written in the twentieth century. Some of these works have become segues to other works merely because they are more exhaustive and have a smattering of Arabic words spread throughout the text, giving the impression of being the work of Arabists. One such is *The Rise of Early Modern Science: Islam, China and the West*, the 1993 book of American sociologist Toby Huff.

Huff constructed a more comprehensive framework for his thesis, which however ultimately makes the familiar claim that there is something inherently flawed in Islam as far as science is concerned; science therefore could not flourish in Islam, and Muslims did not produce a Scientific Revolution. To construct his arguments, Huff identified four factors from within the framework of sociology of science used in the West and applied them both to Islamic scientific tradition and European civilization, to show why the Scientific Revolution took place in Europe and not in the Muslim world. These four factors are as follows: the role of the scientist; the social norms of science; the common elements of scientific communities; and the comparative, historical, and civilizational study of science. The first factor is identified with the work of Joseph Ben-David; the second is an extension of Max Weber's 1949 book *The Methodology of the Social Sciences*; the third builds on the work of Thomas Kuhn, resulting in the idea of paradigms; the fourth proposes a comprehensive approach that calls "for going beyond Max Weber" and that takes into consideration such works as Joseph Needham's monumental study *Science and Civilization in China* (Huff 1993, 14–16).

Islam, Transmission, and the Decline of Islamic Science   127

Armed with this framework, Huff sets off to study "The problem of Arabic science" in the second chapter of his book. Here is his initial casting:

The problem of Arabic science has at least two dimensions. One concerns the failure of Arabic science to give birth to modern science; the other concerns the apparent decline and retrogression of scientific thought and practice in Arabic-Islamic civilization after the thirteenth century. (Huff 1993, 45)

Judgment is already passed before any arguments and proofs are advanced; thereafter, Huff attempts to show the inherent flaws of Islam by a backwards construction of all the factors needed for the growth of science—all based on the conditions prevailing in Europe during and after the Scientific Revolution. The foregone conclusion is that science can flourish only and only if these conditions are present in a civilization; these conditions were not present in the Islamic civilization; therefore science did not flourish. After this the only remaining task for the sociologist is to prove that these conditions did not prevail in the Islamic civilization because Islam did not allow them.

Evidence for this is culled from sources already in use by Orientalists, including the customary reliance on Goldziher. On the question of the dating of decline, Huff first identifies the thirteenth century, then advances this date by a whole century (Huff 1993, 48). Later he eulogizes "achievements of the Arabic science" in various fields, only to return to revisit Goldziher's views: Arabic science came very close to a modern scientific revolution, but did not go "the last mile" because that "metaphysical transition would have, of course, forced an intellectual break with traditional Islamic cosmology as understood by the religious scholars, the *'ulema'*" (Huff 1993, 60).

The reasons advanced for the failure of "Arabic science" to make the final leap are then postulated in the context of philosophical, religious, and legal social roles in the medieval period. Finally we are given the answer in Goldziherian terms: "In general, the structure of thought and sentiment in medieval Islam was such that the pursuit of the rational or ancient sciences was widely considered to be a tainted enterprise" (Huff 1993, 68).

One can keep citing text after such text *ad nauseam* but this adds nothing to our understanding. A major problem with all such works is their inability to explain why science flourished for so long in Islamic civilization in the presence of the so-called Islamic orthodoxy and all the "internal factors" identified by Huff and others. These "internal factors" were of course already present when the Islamic civilization gave birth to and nourished its scientific tradition. It is unreasonable to think that the Islamic legal system, which came into existence in the seventh century (before the emergence of the scientific tradition), would first allow a scientific tradition to flourish for eight centuries and then suddenly become an impediment to the emergence of a "neutral zone of scientific inquiry in which a singular set of universal standards" could be applied.

Recently there has appeared a new kind of response to such works within the broad field of the history of science. This more reflective response has delivered a final blow to Goldziherism and has the potential to provide us more fruitful tools for understanding the nature of scientific enterprise in Islam. One such work is the 1996 paper on mathematical sciences in Islam from a cultural perspective by the Canadian historian of mathematics, J. L. Berggren. Berggren has expanded the customary approach of a historian of science to examine some of the arguments that Goldziher and his followers have advanced. He states that when we look at cultural factors in the growth of a scientific tradition, a problem arises because "there are cultural factors that condition *our* thought, not the least of which is the fact that we do so as members of a civilization whose mathematical development depended importantly on the contributions of the medieval Islamic civilization" (Berggren 1996, 266). In such studies, judgments passed on the scientific achievements of a previous civilization are invariably based on the developments in modern science. This creates historiographic problems and entails the danger of unconsciously slipping from the historical fact into a Whiggish view of history, as if the final purpose of the cultivation of science in the other civilization was merely to create modern science. "This approach has had two quite opposite, but equally regrettable, results," says Berggren:

The first is a treatment of medieval Islam as a civilization deserving of attention only for its role as a channel through which the great works of the Greeks were carried safely to the eager minds of the European Renaissance. The emphasis falls on the two great periods of translations, that into Arabic in the ninth century and that into Latin in the twelfth and thirteenth centuries, and the developments of the intervening centuries provide little more than a series of anecdotes about one curious result or another that was proved by an occasional great figure.

The second result of this Whiggish attitude is a selective and tendentious reading of medieval Arabic texts to show how Islamic science prefigured that of modern times ... it would be invidious to cite contemporary examples of either of these approaches—and of little interest to cite earlier examples—and I shall only observe that both of these results, which on the surface seem to place such different values on Islamic civilization, should concur in valuing it only insofar as it served ends not its own; this is hardly surprising, since both are motivated by a fundamental interest not in the past but in the present. (Berggren 1996, 266–67)

There is yet another perspective on the question of decline, again from within the history of science. Here the approach is to examine the nature of science in Islamic civilization from within its own framework and see where it could have gone. What possibilities were there for different branches of science within the framework of their fields? Aydin Sayili appended a 24-page appendix to his *The Observatory in Islam* with the title "The Causes of Decline of Scientific Work in Islam" in which he has questioned the basic assumption—implied in most works dealing with the cause of decline of science—that "left to itself, science would progress more or less automatically and that its decline would have to be brought about by definite forces, would have to be imposed by outside factors" (Sayili 1960, 408).

We are left with a puzzle. The enterprise of science in Islamic civilization did decline and eventually disappeared. So far, historians, sociologists, and orientalists have not produced any satisfactory answer to the question that naturally comes to mind—why? This is a compelling question; everyone writing on the subject is obliged to respond. What has been said, however, remains a regurgitation of what has already been said, save a few genuine insights that have yielded only partial answers. This is not a book that can undertake a more detailed inquiry on this question. It is, however, useful to point out that the question of decline of science in Islamic civilization cannot be separated from the overall intellectual, economic, social, and political condition of the Muslim world at the time of decline. As such, this inquiry is actually a subset of a much broader inquiry pertaining to the internal dynamics of Islamic civilization during the seventeenth and the eighteenth centuries.

For our purpose, we can mark the seventeenth century as the dividing line between two very different kinds of Islam and science discourses. The "old discourse" was based on an understanding of science that had emerged from within the Islamic worldview, and even the tensions, heated exchanges, and long-standing debates were reflective of this fact. The "new" Islam and science debates that emerged in the eighteenth century totally transformed the basic terms of discourse. This new discourse is the subject of the next chapter.

Chapter 6

Islam and Modern Science: The Colonial Era (1800–1950)

THE BACKGROUND TO THE EMERGENCE OF A NEW DISCOURSE

Most general accounts of the Islamic scientific tradition as well as those dealing with the relationship between Islam and science focuses on Baghdad, or the "City of Peace," as the round city of Caliph al-Mansur was officially called. This is not without reason; after all, the fabled city, established by the victorious Abbasid Caliph on the site of an ancient village, planned by four eminent architects, and built by 100,000 workers and craftsmen over a period of four years (762 to 766) was the intellectual capital of the world for five centuries. With the unconditional surrender of Caliph al-Mustasim to Hülegü to the Mongol warlord on February 10, 1258, Baghdad lost its glory. The Mongol victory was accompanied by the indiscriminate killing of an estimated 800,000 to two million inhabitants, the destruction of all major public buildings, including shrines, mosques, madrassas, and palaces, and an uprooting of intellectual life. By the end of that century of destruction and decay, however, Islamic scientific tradition had already found a new home in other lands of Islam, such as present-day Turkey, Syria, Egypt, and Iran. By the middle of the next century, two major branches of Mongols—the Golden Horde and the Chagatays of Transoxania—had themselves converted to Islam, providing patronage to scientists and scholars in their own newly built madrassas and observatories.

The reconfiguration of the Muslim world in the post-Mongol era would eventually give rise to three powerful empires: the Safavi (1135–1722), the Indian Timuri (1274–1857), and the Ottoman (1343–1924). These empires extended over a vast area and were extremely rich and resourceful. They patronized the arts and sciences and vied against each other to attract the best minds of the time. In addition, several other smaller states and dynasties supported science. Yet, none of these empires was able to compete with the developments in science that were taking place after the fifteenth century in Europe. When they did realize the enormous military and economic benefits Europeans had reaped with their science and technology, it was already too late; better-trained European armies, equipped with superior weapons, were already knocking at their doors.

The rapidity with which the situation changed for Muslims is evident from the fact that at the beginning of the eighteenth century, the entire Middle East, a large part of Africa, the whole middle belt of Asia, and the Malayan archipelago were under Muslim control, but by the end of that fateful century, a large part of this territory had come under Russian, British, French, Portuguese, and Dutch influence or direct control; by the middle of the nineteenth century, there was nothing left of the power, might, and splendor of the old Muslim world. It is against this background that the new Islam and science discourse must be viewed.

In addition to the developments within the Muslim world, the new Islam and science discourse was influenced by the enormous changes that took place in Europe through the application of newly discovered scientific knowledge. The work of Isaac Newton (1642–1727) made a tremendous impact. Two centuries of European science found a new synthesis: the natural world began to lose its qualitative aspect. Instead of form and matter, the four qualities, and the four elements, equations and numbers now started to gain centerstage. The Newtonian Revolution transformed the nature of science.

Newton had shown in the 1680s that the orbits of the planets are the result of an attractive force between the sun and each planet, thus bringing into science a revolutionary concept: gravity. Conceived as a force that worked in a universal manner, whether the bodies on which it operated were heavenly or not, gravity was one of the main concepts of Newtonian physics. His *Principia*, first published in Latin in 1687, firmly established the mechanical model in which bodies were endowed with mass and subjected to external forces such as gravitational attraction.

Ironically, the Islam and science discourse in the centuries following Newton was influenced not so much by the science of Newton and those who came after him but by technologies developed on the basis of their science. The steam engine invented by James Watt (1736–1819), the hydraulic press invented in 1795 by Joseph Bramah (1748–1814), certain technologies used in the extraction of coal and the purification of metals, and military technologies would be looked upon by Muslims as "wonders of European science."

This whole process of change, which had been taking place in Europe for over two centuries, was brought home for Muslims when Napoleon arrived in Egypt in 1798. His army had superior weapons, better training, and was accompanied by scientists specializing in diverse branches of science and several new technologies. Muslims ascribed their military subjugation to a lack of science and technology, and their leaders told them that their lost glory would be restored as soon as they caught up with Europe in science and technology. Religious scholars used their influence to support this call for the acquisition of science. They could easily produce evidence from Islam's primary sources in support of their rallying cry, as both the Qur'an and the Sunnah are replete with exhortations to believers to acquire knowledge. The Arabic word for knowledge, *ilm*, was now used for European science.

The Ottomans were perhaps the first among Muslims to realize their poverty in technological knowledge. This came about through military defeats. The peace treaty signed on January 26, 1699, at Carlowitz was the end of a long chapter in their history and a fateful beginning to a new century. The loss of a large portion of their empire was not the only factor in this defeat; the entire equation between them and Europe had changed, as became clear in 1718, when they were compelled to sign the treaty of Passarovitz, losing Belgrade. The Ottomans were now convinced of the need for reform. This realization evolved into an entirely different era, the famous "Tulip Age," during which the high culture of Ottoman society developed a craze for European civilization, resulting in the abandonment of traditional patterns of design, architecture, music, painting, poetry, and furniture in favor of European styles. This new sensibility was an early sign of what was to follow. And although Belgrade was regained from the Austrians in 1739, this was a short-lived victory. The imperial center could no longer hold its parts; by 1774 the Ottomans had lost control of the Black Sea to the Russians; in 1783, they annexed the large territory around the Sea of Azov. As a result of these setbacks, Salim III undertook far-reaching military and political reforms in 1792. The acquisition of science and technology were the main features of this reform.

In India, the fatal battles of Plassey (1757) and Buxar (1764) consolidated the British hold over Bengal and paved the way for further conquests in Bihar and Orissa. When the British won the battle against Tipu Sultan, the visionary ruler of Mysore, in the closing year of the eighteenth century, the fate of India was sealed, for Tipu's army was the last real resistance against the colonization of India. In 1813 the British government decided to increase its direct control over the East India Company (EIC) through a new charter, ending the monopoly of the EIC and thus paving the way for the full colonial structure. It was then that modern European science arrived in India, and along with it came a new Islam and science discourse.

This new discourse was contemporaneous with a violent transformation of the Muslim world. Yet, on the eve of this violent transformation, there seemed to be no awareness in the Muslim world of what was just around the corner. In fact, the political leadership of all of *Dar al-Islam* (the land of Islam) and its great population seemed like an ocean sleeping before a storm: only a small segment of the Muslim intelligentsia was aware of what was happening. Their awareness produced a series of reform–renewal movements throughout the traditional Muslim lands, which attempted to reconstruct Islamic thought and Muslim societies. These movements were first and foremost religious in nature, but their aim was also to revive Islam's formidable tradition of learning and strengthen society. These reform–renewal movements included the movement led by Sidi al-Mukhtar al-Kunti (*ca*. 1750–1811) in the Sahara and the two West African movements of Uthman Dan Fodio and Shaykh Ahmad of Massina. In the Indian subcontinent, the eighteenth century witnessed a major reform–renewal movement led by Shah Wali Allah al-Dihlawi (1702–1762). This internal process of reform and renewal was, however, cut short by the invasion and colonization of these societies by the European powers.

These movements were ineffectual as far as the general decay was concerned. Of course,

in the fifty some generations of Muslim history, three or four generations hardly suffice to indicate any long-term trend. Yet the depression of Islamic social and cultural life in the late seventeenth and eighteenth centuries does stand out in retrospect. This is so chiefly in the light of what followed. With the nineteenth century came the utter collapse of the strong Muslim posture in the world: that nothing was done in the eighteenth century to forestall this smacks of inexplicable weakness or folly. But the sense that there was a depression also reflects the actualities of the Muslim lands in the eighteenth century itself. (Hodgson 1974, vol. 3, 134)

What actually happened in the three powerful empires (the Safavi, the Indian Timuri, and the Ottoman) that emerged after the great realignment of the Muslim world in the post-Mongol era is akin to the slow growth of a cancer that remains undiagnosed until it has spread throughout the body. When the rot was detected, it was already too late.

What was decisive in Muslim lands at this time was especially one feature: the West's tremendous expansion of commercial power ... by the latter half of the [eighteenth] century, decay was becoming rout in the Ottoman, Safavi, and Indo-Timuri domains ... by the end of the century, the accumulated strains in the social structure of Islamdom called for radical new adjustments, which did supervene then with the forthright establishment of Western world hegemony. (Hodgson 1974, vol. 3, 137)

One of the most important changes to take place in the colonized Muslim world at this time is related to the status of the Arabic language. During the previous millennium, Arabic had remained the main language of intellectual discourse, encompassing religion as well as the sciences. Even in lands where it was not the common vernacular, all major works were written in Arabic. This shared language of discourse had preserved an internal link with the traditional knowledge of religion. It also served as a link for social and economic transactions. An Indian Muslim could go to Cairo, Baghdad, or Makkah and freely communicate with scholars there in a language not foreign to either of them. This allowed the sharing of traditional terminology, metaphors, and parables, as well as ancestral wisdom and teachings. The colonial rulers replaced Arabic with their own languages in occupied lands. In the Ottoman empire, Arabic was replaced by the Turks themselves as part of a modernization drive. Thus, within a short span of time, where Arabic was not the usual spoken language it became a foreign language. This not only destroyed the means of communication among Muslim scholars but in those countries where Arabic was not commonly spoken, it made the Qur'an and the vast corpus of traditional knowledge inaccessible even to the educated classes. This had an enormous impact on the making of the new Islam and science discourse.

NEW RULERS AND NEW SCIENCE

During the eighteenth and the nineteenth centuries Muslim rulers perceived science as a means of power. This was a direct result of the loss of their political power. The arrival of Napoleon in Egypt was a turning point. After Napoleon's army was driven out of Egypt and Muhammad Ali (1769–1849) took control, one of Ali's most important goals was to modernize Egypt by acquiring modern science. He set up new schools and training colleges where modern science was taught by foreigners hired for inflated salaries. Between 1825 and 1836 his government founded military and naval institutions for training officers and soldiers in the various military professions on modern lines, including some disciplines of modern science.

These measures produced no science. The human resources Muhammad Ali amassed were simply not ready for the kind of science (and technology) he was attempting to introduce. It was like attempting to have a body without a backbone. "The pupils recruited for schools—even when the problem of language was somehow solved—had no idea how to learn a science based on its everyday procedures directly on individual, experimental, innovative inquiry ... [the students] had learned to memorize ancient books, [but] memorization of an engineering textbook could not make an engineer" (Hodgson 1974, vol. 3, 219). In the end, all that Muhammad Ali could do was rupture old traditions without succeeding in establishing a new system.

During the 1830s he turned all his military might against the Ottomans. He established links with the British and the French, leading to the great power game of the Middle East and ultimately to the colonization of the entire region.

Similar power struggles marked the disintegration of the Indian empire and resulted in the colonization of the Indian subcontinent by the British. The process began in 1498 with the arrival of Vasco da Gama (1469–1524) in India via a new sea route. This discovery tremendously changed the fortunes of the European economy and, later, its political power. Accelerated by the opening of new sea routes, commerce between Europe and India—then ruled by the Mughals—increased exponentially. This economic activity eventually led to the arrival of merchants and missionaries in India. They began to interfere in local politics, and within two centuries of Vasco da Gama's arrival in India, the balance of power had shifted in their favor. By the beginning of the nineteenth century the vast subcontinent was in the hands of Britain. In 1857 it became the brightest jewel in the British crown. It was in the backdrop of these developments that a new educational system, based on Western educational curricula, was implanted. At first it met fierce resistance but slowly gained popularity. The missionary schools were welcomed by the elite, for they offered Western-style education that, in turn, was the ladder to economic and social mobility. During the course of that century, a fundamental change in the makeup of Muslim society had been accomplished.

The Safavids were attacked by Afghans, who occupied Isfahan in 1722 but failed to establish their power; all they did was leave behind a shattered and weakened empire. Eventually, Nadir Khan, a talented military general, reorganized the Safavid army and expelled the Afghans. He was, however, not content with that feat; he declared himself the ruler of Persia as Nadir Shah. Thereafter he set on a course of destroying the neighboring Muslim empires. He fought with the Ottomans in 1730, attacked India in 1739, and sacked Delhi. To the north, he attacked the chief Uzbeg capitals along the Zarafshan and Oxus rivers. Nothing was rebuilt in the wake of this wave of destruction. Nadir Shah, like Muhammad Ali, found a vacuous region and ascended to its kingship merely on the basis of his personal talent. When he was killed in 1747, Karim Khan Zand, a general from Shiraz, tried to restore the Safavid empire but failed. He ruled over a small region in his own name until 1779. After this there arose another tribal power, the Qajar, who consolidated their hold over the entire region and founded an empire that lasted almost one hundred and fifty years, until a military *coup d'état* brought Reza Khan to power in 1921. In 1925 he extracted constitutional kingship from the Assembly in his own name, thus establishing the Pahlavi dynasty, which was overthrown in 1979 in a popular uprising.

In Ottoman Turkey, the Grand Vizier Damad Ibrahim Pasha (from 1718 to 1730) introduced a number of measures aimed at the modernization of Turkey, including naval reforms and the use of the printing press. Scientific geography became part of the curriculum of military schools. New training centers and schools were introduced on European models, often with the help of European experts and converts. There were some visible results. A two-arc quadrant was invented by Mehmed Said, the son of the Mufti of Anatolia, for use by artillerymen; treatises on trigonometry, new works on medicine, as well as translations of certain European scientific and philosophical writings started to appear. The new naval schools of mathematics established in 1773 produced a new corps of engineers and artillery. Among all the eighteenth-century changes in Turkey, the introduction and acceptance of printing is perhaps the most important. Already in the fifteenth century, certain Jewish refugees, fleeing Spain, had set up printing presses in Constantinople in 1493 or 1494; this was followed by similar presses in other cities. But Arabic and Turkish texts could not be printed due to a ban that was not lifted until July 5, 1727. The first book published from a fully Turkish press was the dictionary of Vankuli, published in February 1729 (Inalcik and Quataert 1994, vol. 2, 637–724).

The nineteenth century witnessed a series of reforms, all aimed at modernizing Turkey. New schools were opened. European languages were taught; scientific texts were translated into Turkish; the first modern census and survey was carried out in 1831. In 1845, a commission of seven eminent men was made in charge of proposing educational reforms. Its report, submitted in August 1846, called for the establishment of an Ottoman state university, a network of primary and secondary schools, and a permanent Council of Public Instruction.

The Ottoman leaders of the nineteenth century were obsessed with the idea of reform, but no matter how many reforms they carried out the economic, political, and social difficulties of the empire remained unsolved. This process of reform produced some critical minds bold enough to criticize their own rulers and society and call for basic change, but it did not reform the society.

With the major defeats suffered by the Ottomans during World War I and after the signing of the treaty of Sèvres on August 10, 1920, the Sultans had lost much public support. This led the way for the emergence of Mustafa Kemal Paşa, later known as Kemal Atatürk (1881–1938), who launched a war of liberation to save Turkey from falling into the hands of European powers. On March 3, 1924, he abolished the Caliphate, banished all members of the Ottoman Sultanate from the Turkish territory, and Turkey became a secular republic. This was followed by the abolition of the Islamic institutions of the country: the office of the Shaykh al-Islam; the Ministry of Religious Law, religious schools and colleges; Islamic courts. These actions created a great deal of resentment and public outcry. Demonstrations followed. Spontaneous armed groups emerged. Now the Republic established in the name of freedom resorted to military action. Kurds, a very large number of citizens of the eastern provinces, leading religious scholars, old nobility, and even young Turks who had been part of the Kemalist revolution but who now differed with Kemal's policies were subjected to atrocities. For Mustafa Kemal and his associates, civilization meant European civilization, as one of Kemal's close associates, Abdullah Cevdet, wrote in 1911 (Lewis 1961, 267). For them, everything related to Islam meant backwardness. They banned traditional dress. The fez, the Turkish hat which had remained a sign of nobility for centuries, was deemed "an emblem of ignorance, negligence, fanaticism, and hatred of progress and civilization" (Lewis 1961, 268). In another speech, Mustafa Kemal declared: "I do not leave any scripture, any dogma, any frozen and ossified rule as my legacy in ideas. My legacy is science and reason."

This dissociation from Islam by the new rulers of Turkey was not shared by the masses, who continued to practice their religion. But the state had gained enormous powers and ruthlessly curbed any public display of religious loyalties. It expunged Islam from the curriculum and imported a large amount of "science" from Europe. Islam was now perceived as the greatest enemy of science, which was seen as the only means of progress and civilization. This official perspective on the relationship between Islam and science was to be propagated vigorously throughout Turkey.

This is the social, political, and intellectual backdrop against which the new Islam and science discourse is to be explored. This new discourse took shape alongside the violent changes just described and was directly influenced by them. The new Islam and science discourse that emerged in the post-1850 era is so different from the old discourse that a modern Muslim scientist is unlikely to find any resonance with men who were the most learned scholars and scientists of the period from the eighth to the sixteenth centuries: Ibn Sina, al-Biruni, and Ibn al-Haytham are not household names in the Muslim world; even an educated Muslim today knows very little about their work and the kind of understanding they had about nature. What is most remarkable in this is a total break with the past, and the most important vehicle of this transformation is education: since their colonization, Muslims have learned to forget the intellectual tradition that produced men like al-Khwarizmi, Ibn Sina, and Ibn al-Shatir. This tradition was violently plucked out of Muslim lands, leaving them bereft of historical depth. Muslim societies have become victim of a cultural schizophrenia in which the past appears as a ghost to be exorcised.

The new Islam and science discourse that emerged in the Muslim world in the nineteenth century has two main strands: (i) a dominant and popular discourse in which Islam is seen as a justifier for the acquisition of modern science and (ii) a distinct but minority position that attempts to maintain a close link with the past, and views the relationship between Islam and modern science in terms similar to the discourse prior to the nineteenth century.

These two aspects of the discourse have developed side by side. Both have been influenced by the manner in which modern science arrived in the Muslim world during the eighteenth and the nineteenth centuries, especially when compared to its emergence in Europe during the sixteenth and the seventeenth centuries. The latter was an organic development, emerging from and deeply entrenched in the matrix of European civilization, with its roots going back to Antiquity. The arrival of modern science in Muslim lands, on the other hand, is akin to the transplantation of an imported plant into an artificially created environment. The growth of modern science in Europe was a natural process linked to the social, economic, political, and intellectual currents of Europe; the growth, even survival, of modern science in the Muslim lands depended (and still depends) upon the maintenance of the artificial environment under which the implant can survive. The former has naturally remained in perpetual contact with all aspects of the civilization that gave birth to it; the latter remains an odd entity in the Muslim world, which received the implant while it was under colonial occupation.

What is "odd" about this process is, however, not immediately apparent. In fact, contrary to this observation, most Muslims would eagerly claim that modern science is really nothing but a refined version of "our own science," which Europe developed and returned. This view is perpetuated through hundreds of websites, conferences sponsored by governments and rulers, and through popular publications. These popular perceptions deeply influence the new Islam and science discourse.

That modern science is an odd entity in the Muslim lands requires explanation. What is meant by its "oddity" has two aspects: (i) the manner of its arrival and (ii) a state of permanent paralysis in which this implant has lived ever since its arrival. Both the arrival and survival of modern science is more by legislative acts, decrees, and proclamations of sultans, charlatan generals, self-appointed presidents, and ministers of science and technology than through the emergence of able scientists, laboratories equipped with instruments, and libraries filled with research papers. It is strange, for example, that the Organization of Islamic Conference, with its headquarters in Jeddah, has a permanent "Committee on Scientific and Technological Cooperation" with the expressed aim of promoting science and technology in the Muslim world, yet it has no scientific institutions, laboratories, or journals. Almost all Muslim states have ministries and ministers of science and technology, who ceaselessly issue statements on the need to acquire modern science, but none of these fifty-seven Muslim states produce any science worth its name and most able Muslim scientists live outside these states.

The following section explores various facets of the two main strands of the new Islam and science discourse. While the first strand is discussed under the simplified subheading of "Islam as a Justifier for Science," it is more complex than that, for it has given birth to a new kind of *tafsir* (commentary) literature—the *tafsir al-ilmi*, the scientific *tafsir* of the Qur'an—which attempts to prove scientific facts and theories from the verses of the Qur'an. This first strand of the new discourse is mixed with many challenges of modernity that Muslim societies face, a new agenda for education and various social and political events in the Muslim world. Rather than being an academic discourse, it has dimensions that often spill over into politics. In short, it is a highly complex mixture with the desire to justify acquisition of science through religious rhetoric, political and social awareness of the need for reform, and various other aspects of the Muslim world as it came into direct contact with Europe through colonization.

THE MAKING OF THE NEW ISLAM AND SCIENCE DISCOURSE

Prominent among those who shaped the new Islam and science discourse in the nineteenth and early twentieth centuries are the Indian scholar and reformer Sayyid Ahmad Khan (1817–1898); Jamal al-Din al-Afghani (1838/9–1897), whom we have already met; his contemporary Egyptian scholar Muhammad Abduh (*ca.* 1850–1905); his Syrian student and later colleague Rashid Rida (1865–1935); the Turkish writer Namik Kemal (1840–1935) and his countryman Badiuzzeman Said Nursi (1877–1960), founder of an important intellectual and religious movement in Turkey. The nineteenth-century Iranian philosophers who wrote in the grand tradition of Islamic philosophy—which had found its greatest synthesis in the person of Mulla Sadra (1571–1640)—include Sayyid Muhammad Husayn Tabataba'i (1892–1981), the author of a major commentary on the Qur'an, *Al-Mizan fi Tafsir al-Qur'an*, Murtaza Mutahari (1920–1979), and Ayatollah Hasan-Zade Aamuli (1929—). Let us explore the new discourse in some depth.

ISLAM AS A JUSTIFIER FOR SCIENCE

In the wake of the arrival of European armies in Muslim lands, the most dominant approach to science rests on the perception that Islam is a religion that supports the acquisition of knowledge: modern science is knowledge; acquisition of knowledge is an obligation of all believers; Muslims must, therefore, acquire science. This knowledge, it is further argued, cannot contradict Islam, for science studies the Work of God and the Qur'an is the Word of God, and there can be no contradiction between the two. First used by Muslim reformers in the nineteenth century and thereafter constantly promoted, this call to "acquire science" has remained unsuccessful. Most of the champions of this rallying cry were (and are) neither scientists nor religious scholars but they are reformers, who considered the enterprise of modern science a means to power and progress. They saw Muslims in need of both, and hence they used Islam to justify their agenda.

The reformers' Islam and science discourse often uses material from Christianity and science debates, including the formulation drawing the link between the "Work of God" (nature) and the "Word of God" (scripture). What they truly desire, however, is fundamentally neither science nor the study of nature; they want to bring the Muslim world out of its state of dependence and decay. Among the early leaders of this approach to Islam and science was Sayyid Ahmad Khan (1817–1898), who was active in politics and education in the Indian subcontinent under British rule. He was to leave a deep mark on the new Islam and science discourse through his writings and by influencing at least two generations of Muslims who studied at the educational institutions he founded.

A great admirer of the English, Khan developed his idea of a modern Islam and a modern Muslim polity while living under British rule. His early writings contain numerous oaths of loyalty to the British rulers of India. He had some of these translated into English and sent copies to high officials of the British authorities in India. He published *An Account of the Loyal Mohamadans of India* (*Risalah Khair Khawahan Musalman*) in 1860, in which he claimed that the Indian Muslims were the most loyal subjects of the British Raj because of their kindred disposition and because of the principles of their religion. He was motivated to show Muslim loyalty to the British because of the persecution of Muslims after the 1857 attempt at liberation known as the Great Mutiny. He appended to one of his works a *fatwa* (religious decree) issued by none other than the Mufti of Makkah, Jamal ibn Abd Allah Umar al-Hanafi, which declared that "as long as some of the peculiar observances of Islam prevail in [India], it is *Dar al-Islam* (Land of Islam)." This was to counter the religious decrees that had been issued by many Indian religious scholars stating that the

The Indian reformer Sayyid Ahmad Khan (1817–1898) attempted to harmonize Islam and modern science by reinterpreting the Qur'an. © Center for Islam and Science.

Indian subcontinent had become a *Dar al-Harb*, the land of war (and thus where military action was religiously legitimate). This political move was favorably received in the ruling circles, and Khan was accepted as an important link between the British and the Indian Muslims. His efforts were directed toward educational reforms. He concluded that Muslims were backward because they lacked modern education. He equated modern education to modern science. He established a Scientific Society, convening its first meeting on January 9, 1864, for four specific goals:

(i) to translate into such languages as may be in common use among the people those works on arts and sciences that, being in English or other European languages, are not intelligible to the natives;
(ii) to search for and publish rare and valuable oriental works (no religious work will come under the notice of the Society);
(iii) to publish, when the Society thinks it desirable, any [periodical] which may be calculated to improve the native mind;
(iv) to have delivered in their meetings lectures on scientific or other useful subjects, illustrated, when possible, by scientific instruments. (Malik 1980)

In 1867, Ahmad Khan and the Society moved to Aligarh, where he procured a piece of land from the British government to establish an experimental farm. The Duke of Argyll, the Secretary of State for India, became the Patron of the Society and Lt. Governor of the N.W. Province its Vice-Patron. Ahmad Khan was the secretary of the Society as well as member of the Directing Council and the Executive Council.

Khan was an ardent believer in the utility of modern science, a devout Muslim, an educator, and a man with a cause. His personal influence grew rapidly. His educational efforts were to change the course of Muslim education in the Indian subcontinent, and his ideas were to have a major impact on the subsequent history of India. He dedicated his life to the uplifting of Muslims in India. He devoted all his energies and a portion of his personal income to the Society he established for the promotion of science. Eventually, he started to receive small sums from like-minded Muslims and from non-Muslim philanthropists, who saw in him a man of vision. On May 10, 1866, he established another organization: The Aligarh British Indian Association to Promote Scientific Education. Within two years the Association was in a position to assist persons traveling to Europe for educational and scientific purposes, but not many Muslims were interested in such trips at that time. Khan himself had never been to England, but he had been elected an honorary Fellow of the Royal Asiatic Society of London in 1864. He now decided to go to England to see for himself the ways of the British in their homeland. Khan went with his two sons, Sayyid Hamid and Sayyid Mahmud. They left India on the first of April 1869; to pay for this trip, Khan had to mortgage his ancestral house in Delhi and borrow 10,000 rupees (Panipati 1993, 3–4). While in England, Khan was awarded the title of the Companion of the Star of India by the Queen. His stay in England convinced him of the superiority of the British. "Without flattering the English," he wrote in his travelogue, "I can truly say that the natives of India, high and low, merchants and petty shopkeepers, educated and illiterate, when contrasted with the English in education, manners, and uprightness, are like a dirty animal is to an able and handsome man" (Khan 1961, 184).

Khan was, first and foremost, an Indian Muslim living at a time when the entire Muslim world was in a state of deep slumber. He wanted to wake them up. He wanted them to acquire modern science and be among the honorable nations of the world. He admired the English for their science and learning, but when he read William Muir's biography of Prophet Muhammad, it "burned [his] heart ... its bigotry and injustice cut [his] heart to pieces" (Panipati 1993, 431). He decided to write a refutation in the form of his own biography of the Prophet. He felt the need to do so not merely for academic and religious reasons but also because his own genealogy connected him to the Prophet. His book was finished in February 1870 and published by Trubner & Co., London, the same year. Khan returned home on October 2, 1870.

During his stay in England he visited the Universities of Oxford and Cambridge and a few private schools, including Eaton and Harrow; these would later serve as models for his own Muhammadan Anglo-Oriental College, established seven years after his return to India. (In 1920 the College became Aligarh Muslim University, now one of the oldest universities of India.) In 1886, he started yet another institution, "The Muhammadan Educational Conference," which organized various conferences in major cities for several years.

Khan's preoccupation with modern science and Islam was so intense that he wanted to lay the foundation of a new science of *Kalam*. This new science would either combat the bases of modern sciences or demonstrate that they conformed to the articles of Islam. Personally, however, he was convinced that Islam and modern science were perfectly aligned, and that all that was needed was reinterpretation to show that the work of God (nature and its laws) was in conformity with the Word of God (the Qur'an).

To prove his views, Khan decided to write a new commentary on the Qur'an. This was a feverish effort that began in 1879. Khan knew he was running out of time, and started to publish his commentary as it was being written. When he died in 1898, this most important work of his life was still incomplete. His work was severely criticized by religious scholars and other Muslim intellectuals, who pointed out his lack of training in Islamic sciences and his inability to use Arabic sources; his zeal to show the agreement between the Word and Work of God earned the pejorative title of *Néchari* ("naturalist"). The minimum qualifications for writing a commentary on the Qur'an were a command over Arabic, a sound knowledge of the sayings of the Prophet, and a thorough grounding in the science of interpretation; Khan lacked all of these. In addition, he did not have an understanding of modern science. His unfinished commentary attempted to rationalize all aspects of the Qur'an that could not be proved by modern scientific methods. These included matters such as the nature and impact of supplications, which he tried to explain as psychological phenomena. Khan however was not alone in making such an effort, as we will see in the next section.

By the time he died, Khan was regarded as the most influential and respected leader of the Indian Muslim community. He had become the intellectual leader of a new generation of Indian Muslims who went to England for "higher education." He was a loyal subject of the British Raj and was considered an ally by the colonial rulers. He was nominated as a member of the Vice Regal Legislative Council in 1878; ten years later, he was knighted as Knight Commander of the Star of India. In 1889 he was awarded an honorary degree from the University of Edinburgh. Khan's impact on the making of a new Islam and science discourse in the Indian

Jamal al-Din al-Afghani (1838/9–1897) was the most important Muslim intellectual of the nineteenth century. His work on Islam and science influenced many subsequent writers. © Center for Islam and Science.

subcontinent can hardly be overstressed. He was not only a thinker but also a practical man who set up institutions that influenced, and continue to influence, the course of education, intellectual thought, and discourse on Islam and science. He was convinced that Muslims need to acquire modern science. This argument gained considerable currency and is still used by many thinkers and rulers throughout the Muslim world. His views on the harmony between Islam and science were shared by many reformers all over the Muslim world. His naturalistic explanations of the Qur'an were, however, attacked by many religious scholars as well as other thinkers. One such scholar, reformer, and revolutionary of sorts was Jamal al-Din al-Afghani, who also played a major role in the development of the new Islam and science discourse in the nineteenth century. He arrived in India in 1879 just in time to write a major rebuttal of Khan's *Nécheri* views.

There is considerable uncertainty about al-Afghani's place of birth and childhood years. He is said to have received his early education in a religious school near Kabul (Afghanistan), Qazwin (Iran), or Tehran (Iran). He went to India in 1855/6, shortly before the failed uprising against the

British and experienced first-hand, in the repercussions, the cruelty of British rule in India. This visit left a deep mark on the twenty-year-old man who would rise to become the most distinct intellectual voice of the colonized Muslims during the last quarter of the nineteenth century. From India, al-Afghani went to Makkah by land, stopping in various Muslim lands on his way. He performed Hajj and returned to Afghanistan by way of Iraq and Iran, where he became an advisor to the ruler of Afghanistan, Amir Dost Muhammad Khan. He arrived in India in 1879 for a second visit, this time to stay for three years. He spent most of his time in Hyderabad, a semi-independent princely state and the cultural center of India at that time. This is where he wrote his Persian refutation of Ahmad Khan's *Nécheri* (naturalistic) ideas, *The Truth of the Néchari Religion and an Explanation of the Necharis* (1881). Five years later, it was co-translated into Arabic by one of al-Afghani's Egyptian students and fellow reformer, Muhammad Abduh. The translation was published in Beirut with a shorter title, *ar-Radd ala ad-Dahriyyin (Refutation of the Materialists)*, in 1886. The original Persian became popular and was reprinted in the year of its publication (1881) from Bombay. There was also an Urdu translation under its original Persian title, published in 1883 in Calcutta. These translations spread al-Afghani's ideas throughout the Muslim world, while English and French translations of his writings carried al-Afghani's ideas to other parts of the world.

In his response, al-Afghani considered *nécheris* and materialists the "deniers of divinity" who "believed that nothing exists except *matiére* (matter)." He included Darwinism in his response and traced its origins to Epicurus. "If one asked him," he wrote about Darwin,

> why the fish of Lake Aral and the Caspian Sea, although they share the same food and drink and compete in the same arena, have developed different forms—what answer could he give except to bite his tongue ... only the imperfect resemblance between man and monkey has cast this unfortunate man into the desert of fantasies? (Keddie 1968, 136)

Afghani rejected the ideas of those materialists who attributed the cause of all changes in the composition of the heavens and earth to "matter, force and intelligence." He considered these ideas to be "corrupt." He accused the materialists of undermining the very foundations of human society by destroying the "castle of happiness" based on religious beliefs. Al-Afghani enumerated "the three qualities that have been produced in peoples and nations from the most ancient times because of religion" as being (i) the modesty of the soul, which prevents them from committing acts that would cause foulness and disgrace; (ii) trustworthiness; and (iii) truthfulness and honesty (Keddie 1968, 146–47).

Islam and Modern Science: The Colonial Era (1800–1950)   147

These he considered the "foundations of stability of human existence," which "the deniers of divinity, the *neicheris*, in whatever age they showed themselves and among whatever people they appeared," tried to destroy.

> They said that man is like other animals, and has no distinction over the beasts... with this belief, they opened the gates of bestiality... and facilitated for man the perpetration of shameful deeds and offensive acts, and removed the stigma from savagery and ferocity. Then they explained that there is no life aside from this life, and that man is like a plant that grows in spring and dries up in the summer, returning to the soil... because of this false opinion, they gave currency to misfortunes of perfidy, treachery, deception, and dishonesty; they exhorted men to mean and vicious acts; and prevented men from discovering truths and traveling toward perfection. (Keddie 1968, 148)

Concluding his refutation, al-Afghani praised religions, especially "the two firm pillars—belief in a Creator and faith in rewards and punishments" (Keddie 1968, 168). "Among all religions," he said, "we find no religion resting on such firm and sure foundations as the religion of Islam.... The first pillar of Islam is *Tawhid*, [which] purifies and cleans off the rust of superstition, the turbidity of fantasies, and the contamination of imagination." Anticipating an objection to his elucidation, he closed his treatise by saying: "If someone says: If Islam is as you say, then why are the Muslims in such a sad condition? I will answer: When they were [truly] Muslims, they were what they were and the world bears witness to their excellence. As for the present, I will content myself with this sacred text: *Verily, God does not change the state of a people until they change themselves inwardly*" (Keddie 1968, 173).

Al-Afghani's movements were closely watched by British intelligence in India. A report by General Superintendent A. S. Lethbridge tells us that he left India in November 1882 (Keddie 1972, 82, n.1); he arrived in Paris at the beginning of 1883 after a brief stay in London. While in Paris he wrote the famous "Answer to Renan" already discussed in Chapter 4.

Renan's lecture, *l'Islamisme et la science* (Islam and Science), and al-Afghani's response are important for understanding the making of the new Islam and science discourse. The former was to set the tone for the European discourse on the new Islam and science nexus, and the latter shows how a leading Muslim intellectual of the nineteenth century viewed the new science and its relationship with Islam. Renan's case for "Islam against science" was built on the basis of the orientalist studies of the previous two centuries and it, in turn, gave birth to Goldziher's influential doctrine (first published in 1916) that posited a supposed "Islamic Orthodoxy" against "foreign sciences." Goldziher's hypothesis, in turn, determined the nature of much of the twentieth-century Western writings on Islam and science.

Renan is, thus, an important player in the making of this discourse. His main point was that "early Islam and the Arabs who professed it were hostile to the scientific and philosophic spirit" and that science and philosophy "had entered the Islamic world only from non-Arab sources" (Keddie 1972, 189–90). Goldziher would, however, change "early Islam" to "Islamic Orthodoxy" to restate Renan's position with a sophisticated layer absent in Renan's quasi-racist lecture.

Renan had sought to prove that there was something inherently wrong with Islam and Arabs in reference to the cultivation of science. In his response, al-Afghani sought to defend Islam by broadening the arguments. He accepted the "warfare model" between religion and philosophy, and blamed all religions for being intolerant and being an obstacle to the development of science and philosophy. With time, he said, all people learn to overcome these obstacles; Islam and Muslims simply have not yet had this time:

Since humanity at its origin did not know the causes of the events that passed under its eyes and the secrets of things, it was perforce led to follow the advice of its teachers and the orders they gave. This obedience was imposed in the name of the Supreme Being to whom the educators attributed all events, without permitting men to discuss its utility or its disadvantages. This is no doubt for man one of the heaviest and most humiliating yokes, as I recognize; but one cannot deny that it is by this religious education, whether, it be Muslim, Christian, or pagan, that all nations have emerged from barbarism and marched toward a more advanced civilization ... If it is true that Muslim religion is an obstacle to the development of sciences, can one affirm that this obstacle will not disappear someday? (Keddie 1968, 182–84)

Al-Afghani's apologetic approach betrays the weight of the previous three centuries of Muslim disgrace. Yet he rests his arguments on past glories he hopes return:

I know all the difficulties that the Muslims will have to surmount to achieve the same degree of civilization, access to the truth with the help of philosophic and scientific methods being forbidden them ... but I know equally that this Muslim and Arab child whose portrait M. Renan traces in such vigorous terms and who, at a later age, became "a fanatic, full of foolish pride in possessing what he believes to be absolute truth," belongs to a race that has marked its passage in the world, not only by fire and blood, but by brilliant sciences, including philosophy (with which, I must recognize, it was unable to live happily for long). (Keddie 1968, 182–84)

Ever since the first formulations of arguments such as Renan's, many Muslim intellectuals have felt obliged to defend their religion against this argument—but only a few have attempted to recast the entire discourse on a different foundation. They also did not challenge the racialist elements in Renan and other writings of the times, for Renan was articulating a view generally held by many Europeans. Renan believed that in the final analysis, for reasons inherent in Semitic languages, the Semites, unlike Indo-Europeans, did not and could not possess either philosophy or science. The Semitic race, he said, is distinguished almost exclusively by its negative features: it possesses neither mythology, nor epic poetry, nor science, nor philosophy, nor fiction, nor plastic arts, nor civil life. For Renan, the Aryans, whatever their origin, define the West and Europe at the same time. In such a context, Renan, who otherwise fought against miracles as a whole, nevertheless retained one: the "Greek Miracle." As for Islamic science, "It is," wrote Renan, "a reflection of Greece, combined with Persian and Indian influences; in short, Arabic Science is an Aryan reflection" (Rashed 1994, 337).

Al-Afghani's response is typical of a nineteenth-century Muslim who felt humiliated by the lack of science and (what was perceived as) "progress" in his own lands. In retrospect, his position appears a natural outcome of the social, political, and intellectual climate of the nineteenth century. Al-Afghani had seen with his own eyes the power of modern science during his travels in the Western world and he was acutely conscious of the domination of the Western powers in world affairs. His response was to provide a motif for subsequent developments in the emergence of the new Islam and science discourse.

Sayyid Ahmad Khan and al-Afghani were two very different thinkers. They had different backgrounds, training, education, and religious and intellectual perspectives, yet they both agreed that Muslims need to acquire modern science. Both understood science to be the road to power: "There was, is, and will be no ruler in the world but science," al-Afghani had declared in a lecture in 1882. "It is evident that all wealth and riches are the result of science" (Keddie 1968, 102). Al-Afghani was unable to perceive in modern science any spiritual or cultural matrix. He did not recognize any difference between modern science and that cultivated in Muslim lands prior to the Scientific Revolution. He criticized religious scholars who recognized a profound difference between "Muslim science" and "European science." He felt that the religious scholars

> have not understood that science is that noble thing that has no connection with any nation, and is not distinguished by anything but itself. Rather, everything that is known is known by science, and every nation that becomes renowned becomes renowned through science. Men must be related to science, not science to men. How strange it is that the Muslims study those sciences that are ascribed to Aristotle with the greatest delight, as if Aristotle were one of the pillars of the Muslims. However, if the discussion relates to Galileo, Newton, and Kepler, they

consider them infidels. The father and mother of science is proof, and proof is neither Aristotle nor Galileo. The truth is where there is proof, and those who forbid science and knowledge in the belief that they are safeguarding the Islamic religion are really the enemies of that religion. The Islamic religion is the closest of religions to science and knowledge, and there is no incompatibility between science and knowledge and the foundation of Islamic faith. (Keddie 1972, 104–5)

It is interesting to note that in his defense, al-Afghani sought recourse with the man most accused of "destroying science" in Islam: Abu Hamid al-Ghazali. He quoted al-Ghazali in a lecture "On Teaching and Learning" as having said that "Islam is not incompatible with geometric proofs, philosophical demonstrations, and the laws of nature" and "anyone who claimed so was an ignorant friend of Islam. The harm of this ignorant friend to Islam is greater than the harm of the heretics and enemies of Islam" (Keddie 1972, 107–8).

Al-Afghani's contemporary Turkish nationalist leader and poet Namik Kemal (1840–1888) also wrote a response to Renan. His defense was, however, quite weak. He defended the thesis that "nothing in Islamic doctrine forbade the study of the exact sciences and mathematics," but he used an anti-utilitarian and strongly moralistic–religious" approach and failed to grasp Renan's attack (Mardin 2000, 324). He wanted Renan to explicitly state that by "science" he meant mathematics and natural sciences and, if he were to do so, then Kemal would agree that "Islamic culture had thwarted the growth of science" (Mardin 2000, 325).

Among those who played a major role in the making of the new discourse on Islam and science in the generation following al-Afghani, Namik Kemal, and Ahmad Khan, the most important Turkish scholar is Badiuzzeman Said Nursi (1877–1960). Unlike his countryman Namik Kemal, Said Nursi opposed the secular ideas of Mustafa Kemal. He was exiled to western Anatolia in 1925, along with thousands of other Muslims, when the new nationalist regime started to use brute force to curb opposition. He spent twenty-five years in exile and imprisonment. During these long years, he changed into another man—one to whom he would refer as "the new Nursi." Most of his works were composed in remote regions of Turkey, without any books or references. He was to make a very deep impression on the next generation of Turks and the movement he started remains alive and active. His writings have now been published as *Risale-i Nur*, after remaining in clandestine circulation for decades.

Unlike al-Afghani and Ahmad Khan, Said Nursi had considerable knowledge of modern science, especially of physics. He attempted to show that there could be no dissonance between the Qur'an and the modern physical sciences. He found modern science to be useful in conveying the message of the Qur'an. Nursi's impact on the making of the new discourse was twofold: he set the stage for direct analogies between Qur'anic verses and inventions of modern science, and his profound spiritual insights led many of his countrymen and other Muslims back to their religion against a strong state-sponsored secularization that had all but erased Islam from the Turkish public space. He used his rhetorical skills to awaken Turkish men and women from their slumber by showing that Qur'anic verses allude to inventions of modern science, such as railways and electricity, and that they should pursue the Qur'anic verses to gain access to the knowledge that can be helpful to them in this world (Nursi 1998, 262). Though he attempts to interpret the Qur'anic verses in the style of *tafsir*, he did not write a full *tafsir* of the Qur'an from a scientific point of view. Given the trends in the new science and Islam discourse, however, the development of a "scientific *tafsir*" was the most logical outcome.

THE SCIENTIFIC *TAFSIR*

A common feature of almost all scientific books written by Muslim scientists who lived before the seventeenth century is the customary invocation to God and salutation to the Prophet placed at the beginning of their books. After that they state their purpose in writing. What one does not find in these works is a mixture of science and *tafsir* (commentary on the verses of the Qur'an), especially not in support of the scientific theories and facts being described. No one attempted to show that their science was already present in the Qur'an. This was to change with the emergence of the new discourse on Islam and science during the post–seventeenth-century era, when books started to appear with verses from the Qur'an purporting to confirm scientific theories. In the end, a new kind of Qur'anic exegesis appeared in full bloom: *al-tafsir al-ilmi*, the scientific *tafsir*.

We have already mentioned the incomplete *tafsir* of Ahmad Khan published during 1879–1898. Another work of much greater impact was by an Egyptian physician, Muhammad ibn Ahmad al-Iskandarani, and was published in 1880. His book has a long title: *The Unveiling of the Luminous Secrets of the Qur'an in which are Discussed Celestial Bodies, the Earth, Animals, Plants, and Minerals*. Three years after the publication of this work, al-Iskandrani published another book in 1883 on the *Divine Secrets in the World of Vegetation and Minerals and in the Characteristics of Animals*. In both works, al-Iskandarani explained verses of the Qur'an to prove the presence of specific scientific inventions in the Qur'an. This trend of writing a scientific commentary on the Qur'an became so popular that general surveys on the Qur'anic exegeses had to invent a new category of *tafsir* literature to accommodate this genre. Thus, al-Dhahabi (1965) devoted a full chapter in his important survey, *Tafsir wa Mufassirun* (*Exegesis and Exegetes*) to the scientific exegesis. Likewise, J. M. S. Baljon (1961), Muhammad Iffat al-SharqĀwa (1972), and J. J. G. Jansen (1974) note this kind of *tafsir* in their surveys. After the trend was established by al-Iskandarani, several other works of this kind appeared in the Arab world (al-Dhahabi 1985, vol. 2, 348). As the nineteenth century approached its close, scientific exegesis had carved out a place for itself in the *tafsir* tradition.

Gaining a certain degree of sophistication, this kind of *tafsir* became a common feature of Muslim works on the Qur'an in the early part of the twentieth century. One of the most influential was that of al-Afghani's student Muhammad Abduh, *Tafsir al-Manar*. This exegesis was compiled by his student Rashid Rida from a series of lectures on the Qur'an given by Abduh in Cairo. Rida continued (from Q. 4:125) when Abduh died in 1905. *Tafsir al-Manar* was finally published in 12 volumes in 1927. This was one of the most influential works during the first half of the twentieth century, as can be judged from the number of editions of this work that appeared between 1927 and 1950.

The scientific exegesis of the Qur'an reached a high point in 1931, when a twenty-six-volume *tafsir* was published by Tantawi Jawhari (1870–1940). His *tafsir*, called *al-Jawahir fi Tafsir al-Qur'an al-Karim (Pearls from the Tafsir of the Noble Qur'an)*, appeared with illustrations, drawings, photographs, and tables. In his introduction to the work, Tantawi says that he prayed to God to enable him to interpret the Qur'an in a manner that would include all sciences attained by humans so that Muslims could understand the cosmic sciences. He firmly believed Qur'anic chapters complemented what was being discovered by modern science.

In the due course of time scientific exegesis made its way into the main body of *tafsir* literature, as many religious scholars began to comment on science in relation to the Qur'anic verses. At times, a writer would divide his commentary into several parts, such as explanation of words, linguistic exegesis, and scientific interpretation. A work of this kind is Farid Wajdi's *al-Mushaf al-Mufassar (The Qur'an Interpreted)*, published in Cairo without a date on the printed edition. In his "remarks on verses," Wajdi often inserts scientific explanations with exclamations placed in parentheses: "you read in this verse an unambiguous prediction of things invented in the nineteenth and the twentieth centuries"; or "modern science confirms this literally" (Wajdi n.d., 346, 423). Wajdi's commentary is not exclusively devoted to scientific explanations of the Qur'an, but many other works are. Even the titles of these works are suggestive of the importance granted to the nexus between science and the Qur'an by the writers. Such works include *Mujizat al-Qur'an fi Wasf al-Kainat (The Miracles of the Qur'an in the Cosmos)* by Hanafi Ahmad (1954), later reprinted (1960) as *al-Tafsir al-Ilmi fi Ayat al-Kawniyya (The Scientific Exegesis of the Cosmic Verses)*; *al-Islam wa tibb al-Haditha (Islam and Modern Medicine)* by Ismail Abd al-Aziz; *al-Nazariyya al-Ilmiyya fi'l-Qur'an (Scientific Theories in the Qur'an)* by Matb al-Itimad

(1942); *Creation of the Heavens and the Earth in Six Days in Science and in the Qur'an* by Hasan Atiyyah (1992); and many others.

What is common in all such works is the zeal to show the existence of modern science in the Qur'an. This zeal is coupled with a desire to prove that the Qur'an is the Book of Allah, since it was impossible for anyone to know about this or that scientific fact in seventh-century Arabia. They also use the well-known Qur'anic claim of its inimitability by restating it to mean that the Qur'an is inimitable because it contains precise scientific information that no human could have known in the seventh century and some of which remains unknown to humanity even now.

This has been one of the most exhaustive and methodological phenomena in the new Islam and science discourse; by the end of the twentieth century, all verses of the Qur'an that could have been used to show its scientific content had received attention by scores of zealous writers. Lists of "scientific verses" of the Qur'an have been compiled, verses have been divided according to their relevance to various branches of modern science such as physics, oceanography, geology, cosmology (Qurashi et al. 1987), and counted to be 750 out of a total of 6616 verses of the Qur'an (Tantawi 1931). This activity then started to give birth to secondary literature—books, articles, television productions, and audiovisual and web-based material. The second half of the twentieth century saw the expansion of this kind of literature, and this is discussed in the next chapter.

ISLAM, MUSLIMS, AND DARWINISM

On the afternoon of July 1, 1858, Charles Lyell (1797–1875) and Joseph Dalton Hooker (1817–1911), two friends of a man who had lost faith in the traditional Christian understanding of the creation narrative in Genesis, presented a paper at the meeting of the Linnean Society of London. The paper entitled "On the Tendency of Species to Form Varieties; and on the Perpetuation of Varieties and Species by Natural Means of Selection," was written by their friend, Charles Darwin (1809–1882). No one could anticipate the far-reaching consequences of Darwin's paper that afternoon—not even Darwin himself—but that day has become a landmark of sorts for all subsequent discourse on creation. In popular accounts, Darwin's theory would be perceived as stating that human beings evolved from monkeys. This layman account was to become the most dominant strand in the Muslim discourse on Darwinism. To be historically precise it should be noted that it was not in his first book, *On the Origin of Species by Means of Natural Selection, or the Preservation of Favoured Races in the Struggle for Life*, first published on November 24, 1859, but in his second major work, *The Descent of Man, and Selection in Relation to Sex* (1871), that Darwin presented his evidence for the descent of man from some lower form along with a complete "mechanism" for the "manner of development of man from some lower form" (Darwin 1887, chaps. 1 and 3).

It should also be pointed out that there are certain inherent problems in the discourse on Darwinism; these stem from the absence of a common terminology and understanding of various concepts shared by all participants of the discourse. These problems have arisen because, over the decades, the same terms have been used differently by advocates of different positions, and all kinds of fine-tunings have been applied to Darwin's original theory so that we now have an incredibly vast range of concepts pertaining to evolution—all of which employ similar terms. This confusion has advanced to such an extent that even the basic terms mean different things to different people. For example, "evolution" may refer to *teleological* evolution (a purposeful and designed process) within a theistic context or it may refer to a *dysteleological* evolution (a process devoid of purpose and driven by random selection and chance only). Likewise, "creation" can be understood to mean a whole range of concepts—from a literal Biblical understanding to progressive creation to "young Earth" creationism. The end result of this proliferation of terms and concepts in the Evolution/Creation discourse is its general unintelligibility unless extreme caution is taken with terms. In addition, the diversity of conflicting usage of terms is further complicated by their transportation to different religious traditions. Even within a single religious tradition, the basic terminology of discourse suffers from confusion (Dembski 1998, 9; Lamoureux 1999, 10). The extremely personal nature of the theme itself also makes a dispassionate discourse difficult. After all, we are dealing with intimate beliefs that have profound spiritual and moral consequences. Perhaps this is why the discourse on Darwinism has always "shed heat, not light" (Goldberg 2000, 2). Because of this intensely personal impact of the belief in one or the other theory, the subject matter of this discourse has always remained open, not just since Darwin, and is likely to remain so in the future.

Darwin's reception in the Muslim world was accompanied by these and other confusions as well as problems that have their origin in the Muslim world. In addition, the process of Muslim understanding of Darwinism was complicated by a peculiar historical background, which needs to be kept in mind for a meaningful discourse.

MUSLIM RESPONSES TO DARWINISM

At the time they encountered Darwinism, a vast majority of Muslims lived in colonized lands under European occupation. Their understanding of modern science was poor; their education limited; books were rare;

Islam and Modern Science: The Colonial Era (1800–1950) 155

there were no scientific laboratories; and the great tradition of scientific research had virtually disappeared from their lands. In addition, the whole Muslim world was in a state of internal strife. One additional point to note is the presence of a large number of Christian missionaries in the nineteenth-century Muslim world; in many cases, controversies associated with Darwin's theory first arrived in the Muslim lands through the writings of these missionaries or their peers in Europe. It should also be noted that, given the general conditions of education and scientific research in the Muslim world at that time, no Muslim was able to produce any scientific response to Darwinism. The number of Muslims who had direct access to scientific journals in the middle of the nineteenth century was probably not beyond a few dozen. All refutations and acceptances of Darwinism were thus based on philosophical, religious, and emotional grounds.

For all practical purposes, Darwin did not exist in the Muslim mind until the first quarter of the twentieth century. In the nineteenth century, only a few intellectuals who had some idea of what was being discussed in the public sphere in England or France were aware of Darwin's ideas, and even they did not have a clear understanding of the scientific and philosophical background that had shaped Darwin's research and philosophical outlook. The Christian community in Lebanon, and to a lesser extent in Egypt, was far more advanced in modern Western education than Muslims, and hence early Arabic works on Darwinism were mostly written by Christians who, in turn, became the immediate sources for works by Muslim writers in Syria, Lebanon, and Egypt.

At the time of Darwin's arrival in the Muslim world, Western-style education was only available in institutions established by missionaries. Often various groups of missionaries fought for influence in the Muslim world. In mid–nineteenth-century Syria, for instance, the American Protestants and French Jesuits were fierce rivals; both established educational institutions. The Syrian Protestant College (SPC) and St. Joseph's College (established by the Jesuits), both in Beirut, became the two most important centers of Western education in the region. These were not merely educational institutions; the missionaries understood their vocation as the spreading of the gospel and enlightenment, and scientific education was, thus, part of the larger package. The situation in India was similar. Many colleges established by missionaries during the nineteenth century became the sources of Western influence on education and science. These institutions also became centers of translation out of practical need. In order to teach, these colleges needed material unavailable in local languages. The staff had to create it; these teachers had proficiency in languages and they opted for the easiest way out by translating existing French or English texts into local languages. This gave birth to secondary scientific works in languages spoken in the Muslim world. Books on various branches of science that appeared in Arabic, Hindi, or Urdu as a result of missionary effort were at best of modest standard, but they served the purpose of spreading European scientific ideas in the Muslim world. This is how Darwinism first arrived in the Muslim world.

Let us note that what eventually became known as Darwinism (along with its modified versions, such as neo-Darwinism and evolution) arrived in the Muslim world in installments. It was seen as a phenomenon, something novel, current, and interesting, but nevertheless not close to home. For all practical purposes, the "real event" was never in full view of most nineteenth-century Muslim writers on Darwinism. Many based their views and responses on prior philosophical or faith commitments rather than on Darwin's ideas. Often they recycled what was being said in Europe for or against Darwin's ideas. Their responses to Darwinism were, therefore, shaped by a perceived view of Darwin's ideas, rather than his actual work.

Since books published in the Muslim world at that time sometimes omitted the date of publication, subsequent accounts of the Muslim reception of Darwinism has remained difficult to assess, with different surveys reporting different chronologies of events. Sometimes these accounts do not even distinguish between Muslim and Christian Arab writers and treat them as if they are all Muslim responses. All of these factors have greatly clouded the discourse.

The earliest traceable mention of Darwin's theory in Arabic goes back to a book by Bishara Zalzal published in 1879 from Alexandria, Egypt, with the title *Tanwir al-Adhhan (The Enlightenment of Minds)*, some twenty years after the publication of *The Origin of Species* and eight years after the publication of *The Descent of Man*. This 368-page work was dedicated in both prose and poetry to the Ottoman Sultan Abd al-Hamid, featured a handsome portrait of Lord Cromer as "a typical example of the Anglo-Saxon people[,] and praised him in two lines of Arabic verse" (Mohammad 2000, 246–47). Both the title of the book and the portrait of Cromer are telling signs of Zalzal's a priori commitments. Lord Cromer, let us recall, had arrived in Egypt to take charge of its finances shortly before the publication of the book, just after Britain and France forced the deposition of Khedive Ismail and installed a more compliant successor. Cromer was in Egypt for only six months, but his measures created unrest in the army, leading to the formation of a nationalist government in 1881. This, in turn, led to the occupation of Egypt by Britain and the return of Cromer to Egypt in 1883. He was to remain in Egypt until 1907 as Her and later His Majesty's Agent and Consul-General, purportedly as "adviser" to a nominally autonomous Egyptian government but in reality as the country's de facto ruler.

Darwin also arrived in the Arab world through scientific journals, which mushroomed between 1865 and 1929. The three most important scientific

Islam and Modern Science: The Colonial Era (1800–1950)

journals were *al-Muqtataf* (1876–1952), *al-Hilal* (1892–1930), and *al-Mashriq* (1898–1930). The case of *al-Muqtataf* is representative: while its editors

> and those in it were predominantly Christians, they nevertheless managed to identify themselves with the Muslim community by urging all Arabs to follow the example of Western civilization. Arabs could progress, they argued, if they adopted the proper methods of education. Arab writers in *al-Muqtataf* linked the idea of progress with that of evolution. It is no surprise, therefore, to find that *al-Muqtataf* devoted much of its discussion to different aspects of Darwinism. (Ziadat 1986, 13)

Those who played a major role in the making of Muslim discourse on Darwin's ideas in the Arab world thus include both Muslims and Christians.

Al-Afghani wrote his "Refutation of Materialists" *(Al-Radd ala al-Dahriyyin)* in 1881 while he was in British India. This was translated into Arabic in 1885 by his student Muhammad Abduh. The article is polemic in nature. It asks Darwin to explain the causes of variations of trees and plants of Indian forests. "Darwin would crumble," he wrote, "flabbergasted. He could not have raised his head from the sea of perplexity, had he been asked to explain the variation among the animals of different forms that live in one zone and whose existence in other zones would be difficult" (Ziadat 1986, 86). He cites Darwin's illustration of how the continuous cutting of dogs' tails for centuries would produce a new variety of dogs without tails and asks rhetorically: "Is this wretch deaf to the fact that the Arabs and Jews for several thousand years have practiced circumcision, and despite this until now not a single one of them has been born circumcised?" (Ziadat 1986, 87). In his later life al-Afghani softened his stand, but he remained a firm believer in the special creation of Man.

A more accommodating line was adopted by the Lebanese Shia scholar, Hussein al-Jisr (1845–1909), who authored more than twenty-five books. Al-Jisr was born in Tripoli, Lebanon, and he was the teacher of many prominent Arabs, including Rashid Rida, the editor of the influential journal *Al-Manar*. Al-Jisr's views on Darwin are also formulated in the context of western materialism but he makes efforts to reconcile the theory of evolution with Qur'anic teachings. He quotes Q. 21:30 (*We made every living thing from water. Will they not then believe?*) and then agrees with the theory of evolution. "There is no evidence in the Qur'an," he wrote, "to suggest whether all species, each of which exists by the grace of God, were created all at once or gradually" (Ziadat 1986, 94).

This theme of accommodation was to find fuller expression in the works of Abu al-Majid Muhammad Rida al-Isfahani, a Shia theologian from Karbala (Iraq), who wrote a book in two parts, *Naqd Falsafat Darwin (Critique of Darwin's Philosophy)*, in 1941. Al-Isfahani defended a God-based version of evolution and counted Lamarck, Wallace, Huxley, Spencer, and Darwin among those who believed in God. He referred to the works of Imam Jafar bin Muhammad bin al-Sadiq (especially to his *Kitab al-Tawhid*) and to those of the Ismaili writers known as Ikhwan al-Safa' to point out anatomical similarities found in humans and apes, claiming that Darwin could never provide full treatment of these similarities as could the Ikhwan. But he disputed the embryological similarities between man and other animals. He affirmed that the structural unity of living organisms was a result of heavenly wisdom and not a consequence of blind chance in nature; he also demanded identification of first causes.

In 1924, Haeckel's book on evolution was translated into Arabic by Hassan Hussein, an Egyptian Muslim scholar, as *Fasl al-Maqal fi Falsafat al-Nushu wa-al-Irtiqa (On the Philosophy of Evolution and Progress)*. In his seventy-two-page introduction Hussein agreed with some of the scientific ideas propagated by Haeckel but he refuted all ideas against religion, though he tried to reconcile Islam and science. He insisted on a nonliteral reading of the six-days verses in the Qur'an and claimed that what Darwin was saying was heavenly wisdom (*Hikmah Ilahiyya*).

Four years after the publication of Hussein's book, Ismail Mazhar (1891–1962) translated the first five chapters of Darwin's *The Origin of Species* into Arabic, adding four more chapters in 1928. The complete translation was published in 1964. He had already himself written a book on evolution in 1924. Mazhar is one of the many secularist Arabs of this time who saw nothing of value in his own civilization. He advocated adoption of the scientific method not only in education but also in life. He also published a journal, *al-Usur,* which had as its motto the phrase *Harrir Fikrak,* "Liberate your thought." He claimed that Islamic Law may have been suitable for the Arabs of the seventh century but was totally incompatible with modern Arab society. He was, to no one's surprise, an ardent follower of Mustafa Kemal of Turkey.

Various aspects of the new Islam and science discourse discussed in this section existed side by side with a less vocal but more deeply rooted discourse that sought to view modern science from the perspective of Islamic tradition as it had been shaped in the preceding centuries. This aspect of the new Islam and science discourse remained in the shadow of the other strand until the last quarter of the twentieth century. Its presence during the eighteenth and the nineteenth centuries is, therefore, seldom acknowledged. What distinguishes this strand of discourse from the other attitude is its restrained approach to modern science. A fuller articulation of this view had to wait until the middle of the twentieth century and is discussed in the next chapter.

Chapter 7

Islam and Modern Science: Contemporary Issues

In this final chapter of the book we are concerned with contemporary issues in the relationship between Islam and modern science. Islamic perspectives on modern science began to emerge in the closing years of the nineteenth century, if one takes the debate started by Ernest Renan in Paris in 1883 as a starting point. Jamal al-Din al-Afghani's response to Renan's polemic has already been mentioned in Chapters 4 and 6. Since then the discourse has become much more complex. For a better understanding of the complexities involved in the new discourse we can divide it into two broad categories. The first relates to the emergence of new Islamic perspectives on modern science in the post-1950 era; the second addresses issues that are totally new to the discourse. These new issues have arisen as part of a greater process of change in the Muslim world and a brief overview of these processes, which have ushered the Muslim world into the twenty-first century, will be helpful in understanding these aspects of the Islam and science discourse.

The violent transformation of a colonized polity that had undergone tremendous destruction of centuries-old traditions during the colonization period into some fifty-seven nation-states, which emerged on the world map in rapid succession between the two world wars, has not been an easy process. This alone has left a deep mark on the contemporary Islam and science discourse. More than theoretical issues it is the direct impact of modern science and technology on the Muslim world that has determined the direction of the Islam and science discourse in the post–World War II era. The sheer magnitude of changes to the physical landscape of regions that had witnessed little change for centuries, the sudden appearance of roads, railways, airports, telephones, oil refineries, and the Internet in deserts where until recently only camel riders traveled under the vast star-strewn skies could not but influence the way science and technology were perceived by men and women living in these lands. The arrival of new tools and techniques in a world unfamiliar with the scientific principles that gave birth to them is a process that, as Werner Heisenberg (1901–1976) once remarked, has far-reaching consequences for the culture that imports them—for tools and technologies change the way we live, which in turn changes our relationship with the tools and science behind them:

> One has to remember that every tool carries with it the spirit by which it has been created... In those parts of the world in which modern science has been developed, the primary interest has been directed for a long time toward practical activity, industry and engineering combined with a rational analysis of the outer and inner conditions for such activity. Such people will find it rather easy to cope with the new ideas since they have had time for a slow and gradual adjustment to the modern scientific methods of thinking. In other parts of the world these ideas would be confronted with the religious and philosophical foundations of native culture. (Heisenberg 1958, 28)

For the Muslim world, the post-1950 era has been like a rude awakening from a medieval siesta. It is, however, the sudden ushering into the twentieth century, filled with violence and traumas of unimaginable proportions, that has brought the world's one billion Muslims face to face with challenges the like of which they have never faced in their long history. Most of these new challenges are somehow related to science and technologies. The penetrating reach of modern science and technology, their impact on the environment, their ability to reshape and reconfigure lifestyles, and their control over modes of production—all of these have deeply influenced the Muslim world during the last quarter of the twentieth century. For a Western reader the magnitude of this impact may be hard to understand, but to have an idea of this change we remember that from the environs of their holiest place on earth—the Ka'bah—to the remotest desert in Africa, there is no place where life has not been altered for Muslims because of science and technology produced in the West. It is true that all of humanity has witnessed a fundamental change in the spectrum of life, but this change has often produced corresponding adjustments in societies where modern science and technologies are cultivated. Such a process has not taken place in societies where modern science and technologies are imported. What Heisenberg realized in 1958 was, in fact, merely the tip of the iceberg.

The phenomenal impact of modern science and technology on Islamic civilization has also produced a corresponding impact on the Islam and science discourse, resulting in the emergence of two different kinds of discourses in the post-1950 era. The first has produced new dimensions of that discourse, which first developed in the 1800–1950 era; the second is the appearance of an entirely new kind of discourse on modern science. Before we begin to explore these two facets, it is important to note that the temporal demarcations being used here are not definitive but approximate time periods that mark the appearance of a significant change in the discourse. Likewise, it is important to have a general idea of the attempts that have been made to initiate a scientific tradition in the Muslim world, for these developments are related to the new Islamic discourse on science that has emerged in the post-1950 era.

The liberation movements in the colonized Muslim world were predominantly nationalistic in nature. Many of the men leading these movements had gone to England or France for higher education and returned home to demand rights and freedoms of the kind they had observed in Europe. The colonial powers allowed them to emerge as national leaders and in time transferred political power to them—sometimes reluctantly, sometimes willingly, but in all cases with the satisfaction that nothing could reverse the course they had set during their occupation of the Muslim world. The political, educational, economic, and scientific institutions they had established would continue to control new nation states, which were sometimes carved out of geographical areas where no independent state had ever existed in history (especially in the Middle East, where Syria, Iraq, and Jordan had never been completely independent states). The Arab world, for example, which had existed as a mostly cohesive polity for centuries, was divided into twenty-two nation-states, some of which were simply nonviable as far as their human and material resources were concerned. Borders were drawn through the sand, as it were, and states were carved out in a manner that divided tribes and even families. There are now fifty-seven Muslim states in the world; more than half of these are contained in the same geographical area where only three existed at the dawn of the twentieth century.

All new Muslim states that emerged from colonial bondage embarked upon rapid modernization plans. They sent their best minds to the West to acquire science, they imported technologies, and attempted to look like their former colonial masters in nearly all respects, from governance to social norms. Science and technologies became the most coveted commodities in these new states. Almost all of these states established ministries of science and technology, prepared official policies for raising the level of science education, and diverted considerable sums from their budgets for the establishment of new institutions for scientific research.

But none of these measures produced science or technology in any of the fifty-seven nation-states—at least, not the kind of science and technology that altered the course of human history during the twentieth century. What was produced, and what continues to be produced in these nation-states, is a caricature of Western science. This was inevitable, for the rickety structure of scientific enterprise propped up in these nation states is without a sustaining backbone. What these states have done in terms of their efforts for producing science is like erecting a building without a foundation. They sent thousands of students to Europe and North America to obtain doctorates in various branches of science with the hope that they could jump-start the production of science upon their return. The results could have been expected: armed with their PhDs, these students returned home to find there was no infrastructure to do science. Libraries with the latest journals, laboratories with working instruments, industry in need of scientific and technological solutions to its problems, large pharmaceutical and defense industries (which feed enormous sums of money into the enterprise of science in the West)—none of these features of modern scientific research were present in the Muslim world, to which a substantial number of Western-trained men and women returned in the last quarter of the twentieth century. These were, however, not necessarily scientists but merely men and women who had been awarded degrees by Western universities and who could do little more than repeat what they had done during their stay abroad. Those who could do truly original scientific research realized they would have to return to the West to do what their profession demanded, for conditions at home were simply not suitable. And so many returned to the West, not in the hundreds or thousands but in the hundreds of thousands. These men and women now work in laboratories and universities as far north as the Arctic and as far south as the South Pole. They are part of the Western scientific enterprise—able scientists who have made substantial contributions to modern science.

This is a general view of the context into which the two contemporary strands of the Islam and science discourse have emerged. Let us now explore these.

NEW DIMENSIONS OF THE OLD DISCOURSE

The first strand of the contemporary Islam and science discourse is actually a continuation of what had appeared during the colonial era—a discourse in which Islam is used as a justifier for science. This strand of the Islam and science discourse experienced a boom in the 1980s, when various states pumped their new-found oil wealth into sponsoring institutions for "research on the scientific verses of the Qur'an." For example, a "Commission for Scientific Miracles of Qur'an and Sunnah" was established in Saudi Arabia by the World Muslim League, with six goals and objectives:

(i) To lay down governing rules and methods [for studying] scientific signs in the Holy Qur'an and Sunnah; (ii) To train a leading group of scientists and scholars to consider the scientific phenomena and the cosmic facts in the light of the Holy Qur'an and Sunnah; (iii) To give an Islamic Character to the physical sciences through introducing the conclusion of approved researches into the curricula of the various stages of education; (iv) To explain, without constraint, the accurate meanings of the Qur'anic verses and the Prophet's Traditions relating to Cosmic Sciences, in the light of modern scientific finds, linguistic analysis and purpose of Shariah; (v) To provide Muslim missionaries and mass-media with Dawah; (vi) To publicize the accepted researches in simplified forms to suit the various academic levels and to translate those papers into languages of the Muslim world and the other living languages. (as-Sawi 1992)

The Commission has to date published about twenty books dealing with the "scientific miracles" of the Qur'an in various fields such as embryology, botany, geology, astronomy, and cosmology. It organized five international conferences between 1987 and 2000 in various countries, which hosted splendid ceremonies where Western scientists were invited to receive attention and patronage from princes and other high officials of kingdoms and states. These scientists were asked to comment on specific "scientific verses" of the Qur'an on the basis of science. The result was the emergence of a scientific-hermeneutic approach that generated tremendously popular apologetic material. This material proposed to prove to the world the scientific correctness of the Qur'an on the authority of great Western scientists like so-and-so. In due course, these conferences have covered all verses of the Qur'an that have any relevance to various branches of science such as embryology, geology, and medicine. The audiovisual recordings of these conferences are available on scores of websites and numerous books have been published in various languages that use material from these conferences.

A famous case is that of the Canadian embryologist Keith Moore, who was a regular keynote speaker at such conferences during the 1980s. His textbook on embryology, *The Developing Human*, was published by the Commission with "Islamic Additions: Correlation Studies with Qur'an and *Hadith*" by Abdul Majeed A. Azzindani (Moore 1982). In the foreword to this edition, Moore wrote:

I was astonished by the accuracy of the statements that were recorded in the 7th century AD, before the science of embryology was established. Although I was aware of the glorious history of Muslim scientists in the 10th century AD and of some of their contributions to medicine, I knew nothing about the religious facts and beliefs contained in the Qur'an and Sunnah. It is important for Islamic and other students to understand the meaning of these Qur'anic statements about human development, based on current scientific knowledge. (Moore 1982, 10)

During the Seventh Medical Conference held by the Commission at Dammam, Saudi Arabia, in 1981, Moore said that "it has been a great pleasure for me to help clarify statements in the Qur'an about human development. It is clear to me that these statements must have come to Muhammad from God, because almost all of this knowledge was not discovered until many centuries later. This proves to me that Muhammad must have been a messenger of God." During the question session, when Moore was asked, "Does this mean that you believe that the Qur'an is the word of God?" he replied, "I find no difficulty in accepting this."

Similar state-sponsored programs were initiated in Pakistan, Jordan, and other Muslim countries. A precursor to this was the work of a French physician, Maurice Bucaille, who published his enormously popular book *La Bible, le Coran et la science : Les écritures saintes examinées à la lumière des connaissances modernes* (*The Bible, the Qur'an, and Science: The Holy Scriptures Examined in the Light of Modern Knowledge*) in 1976. Bucaille's book has been translated into every language spoken in the Muslim world and hundreds of websites refer to it. Bucaille, who was the family physician of the Saudi King Faisal, attempted to show that the Qur'an contains scientifically correct information about the creation of the heavens and earth, human reproduction, and certain other aspects of the natural world whereas the Bible does not. His book became the main source for dozens of other secondary works on Islam and science. Bucaille's work is a forerunner to numerous other works that attempt to interpret the Qur'an on the basis of modern scientific knowledge. In all such works, the Qur'anic vocabulary is placed within the framework of modern science and its verses are interpreted to show the existence of "scientifically correct" knowledge in the Qur'an.

The creation of the heavens and earth is a popular theme in this strand of Islam and science, where certain verses of the Qur'an are chosen to demonstrate that the Qur'an contains modern scientific data. The two most often cited verses are Q. 21:30 and Q. 41:11. The former states, *Do the disbelievers not see that the heavens and the earth were joined together, then We clove them asunder and We created every living thing out of water. Will they then not believe?* The latter reads, *God turned toward the heaven and it was smoke*... In the first verse, the two key Arabic words are *ratq* and *fatq*; the former is translated as "fusing or binding together" and the latter as the process of separation. These two key words are then used to support the Big Bang model. Other verses pertaining to creation mention "six days" during which the heavens and the earth and all that is between them were created by God. The six days are shown to mean six indefinite periods of time (Bucaille 1976, 149).

There seems to be no problem with the interpretation of six days as six periods, for the Qur'anic usage supports this, but numerous problems begin to surface when this Qur'anic data is superimposed on specific data arising from modern science. Bucaille chose to interpret "smoke" (*dukhan*), mentioned in verse 11 of chapter 41, as "the predominantly gaseous state of the material that composes [the universe, which] obviously corresponds to the concept of the primary nebula put forward by modern science" (Bucaille 1976, 153). It is this one-to-one correspondence that begins to stretch Qur'anic hermeneutics. The entire enterprise remains conjectural, as no proofs can be found for such an interpretation. As the narrative proceeds, the desire to reveal "science" in the Qur'an makes the task of interpretation even more difficult: "The existence of an intermediate creation between 'the heavens' and 'the earth' expressed in the Qur'an may be compared to the discovery of those bridges of material present outside organized astronomic systems" (Bucaille 1976, 153).

The foregone conclusion of this approach toward the relationship between the Qur'an and science is that

although not all the questions raised by the descriptions in the Qur'an have been completely confirmed by scientific data, there is in any case absolutely no opposition between the data in the Qur'an on the Creation and modern knowledge on the formation of the universe. This fact is worth stressing for the Qur'anic Revelation, whereas it is very obvious that the present-day text of the Old Testament provides data on the same events that are unacceptable from a scientific point of view. (Bucaille 1976, 153–54)

The attention received by Bucaille's book has produced reactions as well. One of the more serious rebuttals came from an expected quarter: a Christian response by William Campbell. *The Qur'an and the Bible in the Light of History and Science* attempted to show the opposite of what Bucaille had set out to prove; it is the Qur'an that has it all wrong, while the Bible is sound (Campbell 1986).

Bucaille was building on the trends in Islam and science discourse already present in the nineteenth century. His contribution became more popular than the work of Egyptian physicians who had embarked upon a similar project in the nineteenth century, perhaps because he was a European who fulfilled a psychological need of Muslims emerging from two centuries of colonization. Whatever their utility, in the final analysis such trends remain polemical and they provide little insight into the nature of the relationship between Islam and modern science.

NEW PERSPECTIVES ON ISLAM AND MODERN SCIENCE

One of the most important developments in the discourse on Islam and modern science owes its existence to the work of a few Muslim thinkers living in the West. Ironically, these new insights into Islam's relationship with modern science have not been received in the traditional Muslim lands with the same kind of enthusiasm with which the work of Maurice Bucaille and Keith Moore was received. This is a telling sign of the intellectual climate of the Muslim world, which forced many leading thinkers to leave their homes and migrate to the West. This westward movement of Muslim intellectuals and scientists is part of the general exodus that has brought millions of Muslims to Europe and North America during the last fifty years.

Muslim presence in Europe and North America is a unique historical development with far-reaching consequences. For Europeans and North Americans, Islam and Muslims are no more two unknown and unknowable mysteries—Muslims have literally become next-door neighbors. This situation promises better relations between various faith communities (a promise yet to be realized) and the Muslim diaspora has produced its unique reflections on Islam, Muslim history, Islamic civilization, and science. In many cases, this scholarship emerging from outside the *Dar al-Islam* (the traditional abode of Islam) is the best available material in a given field; such is definitely the case for the Islam and science discourse. This section provides a brief survey of certain new aspects of the discourse.

A broad classification of the current discourse on Islam and modern science identifies three categories: ethical, epistemological, and ontological/metaphysical views of science.

The ethical/puritanical view of science, which is the most common attitude in the Islamic world, considers modern science to be essentially neutral and objective, dealing with the book of nature as it is, with no philosophical or ideological components attached to it. Such problems as the environmental crisis, positivism, materialism, etc., all of which are related to modern science in one way or another, can be solved by adding an ethical dimension to the practice and teaching of science. The second position, which I call the epistemological view, is concerned primarily with the epistemic status of modern physical sciences, their truth claims, methods of achieving sound knowledge, and function for the society at large. Taking science as a social construction, the epistemic school puts special emphasis on the history and sociology of science. Finally, the ontological/metaphysical view of science marks an interesting shift from the philosophy to the metaphysics of science. Its most important claim lies in its insistence on the analysis of the metaphysical and ontological foundations of modern physical sciences. (Kalin 2002, 47)

Another way of classifying recent developments in the Islam and science discourse is to study it through the description and analysis of positions of major thinkers (Stenberg 1996). Whatever way one chooses to classify the new discourse, ultimately it is dealing with a small body of literature that has emerged during the last half of the twentieth century.

Islam and Modern Science: Contemporary Issues   167

These new aspects of the discourse are intimately connected with the entire range of issues emerging from Islam's encounter with modernity. Muslim thinkers have generally regarded this encounter as the most vital in the history of Islam and they have attempted to find viable Islamic alternatives to Western economic, social, cultural, and educational systems in order to preserve Islamic values. This search for a *modus vivendi* includes a reassessment of modern science and technology from an Islamic perspective. The enterprise of science in the West has emerged from a certain historical background; it is highly linked to other institutions of Western civilization, and notwithstanding its claims to universality it is the product of Western civilization. As such, it is deeply entrenched in a worldview different from Islam. In fact, not only science but all modern knowledge has been deemed to require an epistemological correction. This need created a movement that conceived a program of "Islamization of knowledge." Led by Ismail al-Faruqi (1921–1986), the movement was based on the premise that the root of decline of the Muslim world was the "educational system, bifurcated as it is into two subsystems, one 'modern' and the other 'Islamic'" (al-Faruqi 1982, viii). To redress this "malaise," al-Faruqi sought to unite the two educational systems and to Islamize knowledge. Al-Faruqi's approach to the problem of modern knowledge was based on the realization that the earlier reformers in the Muslim lands had remained unsuccessful in their efforts because they failed to understand the deep roots of modern knowledge. They assumed that

the so-called 'modern' subjects are harmless and can only lend strength to the Muslims. Little did they realize that the alien humanities, social sciences, and indeed the natural sciences as well were facets of an integral view of reality, of life and the world, and of a history that is equally alien to that of Islam. Little did they know of the fine and yet necessary relation that binds the methodologies of these disciplines, their notions of truth and knowledge, to the value system of an alien world. That is why their reforms bore no fruit. (al-Faruqi 1982, viii)

The solution to this "Malaise of the Ummah," as al-Faruqi conceived it, was perceived "in concrete terms, to Islamize the disciplines, or better, to produce university level textbooks recasting some twenty disciplines in accordance wit[h] the Islamic vision" (al-Faruqi 1982, 14). This idea led to the establishment of the International Institute of Islamic Thought (IIIT), which continues to pursue al-Faruqi's vision. Al-Faruqi, however, was not interested in studying the epistemological foundation of modern science, and his plan dealt only with the social sciences.

Al-Faruqi's limited approach to the process of Islamization of knowledge drew attention to the absent content (the natural sciences), and a number of other scholars attempted to formulate pertinent questions regarding Islam's relation with modern science. One such attempt was led by Ziauddin Sardar, a UK-based journalist of Pakistani origin, together with a few other scholars who formed a loose-knit group called "Ijmalis." Sardar's major work on the subject, *Explorations in Islamic Science* (1989), was inspired by developments during the previous decade, which had witnessed a surge of interest in Islam all over the world. Sardar focused on a related subject—the role of science and technology in the development of the Muslim World. During his research, he "visited science institutions and universities in many Muslim countries and was struck by the extent of the discussion on Islam and science" (Sardar 1989, 1). He realized that many working scientists

> felt that there were some problems between their religious ethics and their professional work as scientists. No one actually articulated the problem in any clear way—it was slipped in during complaints about how science is ignored, lack of funding, absence of adequate research facilities and so on. When posed a direct question, most scientists avoided talking about ethics in science or the notion of Islamic science. The explanation offered by a Turkish scientist placed this reluctance in perspective: 'Obviously', he said, 'I have my own opinion on the relationship between science and Islam, but I would not discuss the subject in my office or indeed at any scientific or public gathering. This would be the fastest way to lose the respect of one's colleagues, become isolated and labeled as a fanatic. In fact, such a discussion would mean the end of my scientific career.' (Sardar 1989, 1)

This situation was to change. "In less than five years," Sardar noted, "Muslim scientists were more assertive about their religious and ethical concerns" (Sardar 1989, 2). What changed was an understanding of modern science. The first step toward the evolution of this strand of discourse was a realization by a number of Muslim scientists and thinkers that "while science itself is neutral, it is the attitude by which we approach science that makes it secular or Islamic" (Sardar 1989, 2). Thus, according to Sardar, it was now asserted with increasing emphasis that science is intricately linked with ideology in its emphasis, scale of priorities, and control and direction of research. They observed that science promotes certain patterns of growth and development, as well as a certain ideology. The key phrase in Sardar's formulation is the neutrality of science, a concept that has been seriously challenged by a host of scholars, both Muslim and non-Muslim, in recent years.

Sardar in particular and his associates in Ijmalis in general developed their discourse on the following assumption:

> The purpose of science is not to discover some great objective truth; indeed, reality, whatever it may be and however one perceives it, is too complex, too interwoven,

too multidimensional to be discovered as a single objective truth. The purpose of science, apart from advancing knowledge within ethical bounds, is to solve problems and relieve misery and hardship and improve the physical, material, cultural and spiritual lot of mankind. The altruistic pursuit of pure knowledge for the sake of 'truth' is a con-trick. An associated assumption is that modern science is distinctively Western. All over the globe all significant science is Western in style and method, whatever the pigmentation or language of the scientist. (Sardar 1989, 6)

Working with this main assumption, Sardar then developed a second premise for his exploration:

My second assumption follows from this: Western science is only a science of nature and not *the* science. It is a science making certain assumptions about reality, man, the man-nature relationship, the universe, time, space and so on. It is an embodiment of Western ethos and has its foundation in Western intellectual culture. Different constellations of axioms and assumptions may lead the sciences of two different societies to highly divergent interpretations of reality and the universe, interpretations which may either be spiritual or materialistic according to the predisposition of the society. (Sardar 1989, 6)

Sardar and his associates situated science in the social and utilitarian realms, reducing it to no more than a tool for "solving problems and relieving misery." However, higher science dealing with the structure of physical reality has no immediate utility: Einstein's four papers of 1905 neither relieved misery nor solved problems; they had no impact on the nature of the hardship or physical and material lot of mankind, yet they altered our whole concept of mass, time, motion, and light, leading to the emergence of a new kind of physics. Sardar and others in his group were not blind to this, but their emphasis was on a culture-specific construction through which they could raise certain social issues. They built their discourse on the need for each civilization to produce its own specific kind of science within its own worldview, but the difference between the science of one civilization and another was perceived merely in terms of priorities of research, utility of science, social prestige, and salaries of scientists. They left out the ontological and metaphysical considerations from their sociological discourse. In other words, they built an epistemology of science without any philosophy and ontology.

Sardar's work has insights into the concrete realities of the Muslim world: its social, intellectual, and scientific aspects and the deep chasm that is so characteristic of the contemporary Muslim world. Although his discourse is rich in self-contradictions (perhaps because it lacks any systematic foundation), what he contributed to the making of the new discourse is not unimportant. He perceived the real-life dilemmas of Muslim scientists who "tend to propagate two different sets of values: one that is evident in their professional output and another that they cherish in their personal lives" (Sardar 1989, 24). Sardar attempted to explain this by dividing the knowledge of Muslim scientists into the operational and the nonoperational—in other words, their scientific training and their Islamic values. "Most Muslim scientists, therefore, suffer from an acute schizophrenia, the seeds of which are planted at the beginning of their education" (Sardar 1989, 26). The Western educational system was implanted in the Muslim world during the colonial era and it remains the main source of the schizophrenia mentioned by Sardar.

In a positive construction, Sardar identified three elements of Islamic science: (i) humility; (ii) the recognition of the limitations of scientific method; and (iii) respect for the subject under study. This somewhat ad hoc list of elements can actually be extended to include many other elements, such as reverence for the creation of God, an attitude of care and preservation, and so on, without adding or subtracting anything from the enterprise of modern science. What Sardar and his associates failed to see was the foundations on which the modern enterprise of science emerged. Their discourse was more concerned with deconstructing myths, producing an awareness of the enormous differences between the status of Western scientists and those working in the Muslim world, and vehemently rejecting certain trends in the development of Islamic perspectives on science that were becoming increasingly pronounced during the late 1970s and early 1980s. Among these was the phenomenon of "Bucaillism" already mentioned in the previous section. Although Sardar did not realize, he criticized these trends as "dangerous" and traced their motif back to the psychological need of some Muslims to prove that the Qur'an is "scientific and modern." He took a contemporary pamphlet by Muhammad Jamaluddin El-Fandy, *On Cosmic Verses in the Quran*, to be "one of the earliest" examples of such works, without realizing that this apologetic literature was already the high point of Islam and science discourse in the last two decades of the nineteenth century. Regardless of this historically inaccurate aspect, his criticism was instrumental in intensifying an internal critique of various positions within the Islam and science discourse.

The Ijmalis position seemed important during its heyday but was soon shown to lack solid roots for growth; its strongest advocates, who were mostly freelancers, did not make efforts to sustain their discourse. All three champions of this position, Ziauddin Sardar, Pervez S. Manzoor, and Munawwar Anees, moved to other topics during the 1990s.

The result of inter-Muslim debates on the nature of Islamic and non-Islamic sciences was the maturity of the new discourse on Islam and science during the last quarter of the twentieth century. This process was helped by a number of international conferences and seminars held in various Muslim countries. Two of the most important conferences of this nature were held in Pakistan in 1983 and 1995. At such conferences, a whole range of perspectives on modern science could be stated, debated, and thrashed about, with participants returning to their countries with fresh insights. Through this process the new discourse sifted the important from the unimportant.

THE METAPHYSICAL PERSPECTIVE

How does modern science relate to the concept of *Tawhid*, the heart of Islam which tells us that there is no deity except Allah, the Creator? What are the implications of the subtle metaphysical assumptions of modern science? What are the consequences of these assumptions in terms of our understanding of physical reality? How does this understanding differ from the Islamic understanding of the physical world? How are space, time, and matter understood in Islam and modern science? These and similar questions have informed a different kind of strand in the Islam and science discourse. As opposed to sociological and philosophical studies on modern science from an Islamic perspective, this strand of Islam and science discourse has been built upon a metaphysics whose roots go back to the centuries-old Islamic tradition of reflection on physical reality from the perspective of its ontological dependence on the Creator, its relative position in the overall scheme of creation, and its purpose and ultimate end. While exploring this strand of the contemporary Islamic discourse on science we encounter different terminology that deals with the physical cosmos in terms of its sacredness, its inviolability, its ontological status, and its unfathomable links with the higher realms of existence.

Built on the insights of sages of previous centuries, this strand of Islamic discourse on modern science came into existence through the work of a handful of scholars who are often called "traditionalists" for their links with the living spiritual tradition of Islam. This view places the enterprise of modern science in a metaphysical framework and compares it with the premodern scientific tradition to highlight its main features. The traditional sciences that studied the physical cosmos derived their principles from revelation, the traditionalists argue, whereas modern science derives its principles from human reason. As a result of this foundational difference between modern science and the traditional sciences of nature, modern science has embarked upon the study of the physical cosmos in total disregard to its sacredness, and the results have been devastating for the planet as well as for those who inhabit it. Treating the emergence of modern science as a historical process set in a definite geographical region, this view of science links its emergence with other developments in Europe at the time of the Scientific Revolution. One important aspect of this discourse is its emphasis on symbols and spiritual meanings of the physical entities that are the subject of study in modern science.

The main assumption in this discourse is the teleology of cosmos—a self-evident reality displayed in and built into the very nature of the remarkable order of the cosmos (it is not imparted to it by the observer). This view holds that natural science and data gathered by scientific tools and observations should be examined in the light of metaphysical knowledge derived from revelation. The exponents of this view claim that the sacred aspects of this view are part of all revealed religions and hence part of the *sophia perennis* (perennial wisdom). "By *Philosophia Perennis*—to which should be added the adjective *universalis*—is meant a knowledge which has always been and will always be and which is of the universal character both in the sense of existing among peoples of different climes and epochs and of dealing with universal principles," wrote Seyyed Hossein Nasr, an important representative of this school of thought, in his 1993 work *The Need For a Sacred Science*:

This knowledge which is available to the intellect, is, moreover, contained in the heart of all religions or traditions, and its realization and attainment is possible only through those traditions and by means of methods, rites, symbols, images and other means sanctified by the message from heaven or the Divine which gives birth to each other. The epistemology provided by *sophia perennis* covers an incomparably greater range of possibilities since it opens the way for relating all acts of knowing to the intellect and, finally, to the Divine. (Nasr 1993, 53–54)

Built on the great repository of metaphysical writings of Islamic scholars, this strand of discourse on modern science attained its present form through the pioneering work of a small number of scholars including René Guénon (d. 1951), Frithjof Schuon (d. 1998), Titus Ibrahim Burckhardt (d. 1984), Martin Lings (d. 2005), Charles Le Gai Eaton (b. 1921), and Seyyed Hossein Nasr (b. 1933). At a different level and in his own way, Syed Muhammad Naquib al-Attas (1931—) has also contributed to this discourse. These pioneering studies are producing more fruits through the work of a new generation of writers who have adopted the basic elements of this approach and who continue to expand the work of the previous generation.

This approach is a marked departure from attempts to graft Islamic ethics and values onto modern science through artificial means. Here the discourse is built upon a metaphysical framework of inquiry that constructs a concept of nature according to the primary sources of Islam. Concepts such as hierarchy, interconnectedness, isomorphism, and unity—which are built into the very structure and methodology of traditional sciences of nature—are used here to identify the dissonance of modern science with Islam. Seen from this perspective, modern science appears as an anomaly, "not simply because we have to pay a high price by destroying the natural environment, but because modern science operates within a seriously misguided framework in which everything is reduced to pure quantity and by which modern man is made to think that all of his problems, from transportation to spiritual salvation, can ultimately be solved by further progress in science" (Kalin 2001, 446).

The critics of this approach often construe this discourse as being anti-science, archaic, nostalgic, and impractical. This criticism, however, is the result of partial understanding: one does not find an anti-science attitude in the original work of these writers (i.e., if "anti-science" means a rejection of the need to study and explore the natural world). On the contrary, these writers often reassert the traditional view that the cosmos *must* be studied—because it is a sign of the Creator. What they stress, however, is the framework for this study, which they find unacceptable in modern science.

The enterprise of modern science as it has developed since the seventeenth century is seen by the advocates of this discourse as a disastrous outcome of the loss of the sacred. Not only modern science but the whole outlook of modernity is marked by a loss of the sacred and the ascendancy of the profane. This insight has been brought out most notably in the work of René Guénon and Frithjof Schuon. The enterprise of science cannot be an autonomous undertaking; it is always part of a civilization. In traditional civilizations, sciences were always part of a hierarchy of knowledge that paid attention to the physical world in due proportion, without either negating it or giving it undue importance. With the advent of modern science this hierarchy has been lost, plunging humanity into a state of multiple and deep crises. This process started with the European Renaissance—a time that is understood by the traditionalists as the beginning of the modern dark ages—when "a word rose to honour," a word "which summarized in advance the whole programme of modern civilization: this word is 'humanism'" (Guénon 1942, 25).

Men were indeed concerned to reduce every principle of a higher order, and, one might say symbolically, to turn away from the heavens under the pretext of conquering the earth; the Greeks, whose example they claimed to follow, had never gone so far in this direction, even at the time of their greatest intellectual decadence, and with them utilitarian considerations had at least never claimed the first place, as they were very soon to do with the moderns. Humanism was already the first form of what has subsequently become contemporary laicism; and, owing to its desire to reduce everything to the measure of man as an end in himself, modern civilization has gone downwards step by step until it has ended by sinking to the level of the lowest elements in man and aiming at little more than satisfaction of the needs inherent in the material side of his nature, an aim which is, in any case quite illusory, as it constantly creates more artificial needs than it can satisfy. (Guénon 1942, 25–26)

According to this view, modern science was one of the most important products of this transformation of the Western world. It arose by breaking links with the past. This can be demonstrated by one example.

The term "physics" in its original and etymological sense means precisely the "science of nature" without any qualification; it is therefore the science that deals with the most general laws of "becoming," for "nature" and "becoming" are really synonymous, and it was thus that the Greeks, and notably Aristotle, understood this science. If there are more specialized sciences dealing with the same order of reality, they can only be mere "specifications" of physics for one or another more narrowly defined province... the modern world has subjected the word "physics" to designate exclusively one particular science... this process of specialization arising from the analytical attitude of the mind has been pushed to such a point that those who have undergone its influence are incapable of conceiving of a science dealing with nature in its entirety... (Guénon 1942, 63)

Guénon goes on to describe how the different branches of modern science cannot be said to be the equivalent of the physics of the ancients:

If one were to compare the ancient physics, not with what the moderns call by this name, but with the sum of all the natural sciences as at present constituted—for this is its real equivalent—the first difference to be noticed would be the division that it has undergone into multiple "specialties" which are, so to speak, foreign to one another. However, this is the only the most outward side of the question, and it is not to be supposed that by joining together all these particular sciences an equivalent of the ancient physics would be obtained... The traditional conception attaches all sciences to the principles of which they are the particular applications, and it is this attachment that the modern conception refuses to admit... The modern conception claims to make the sciences independent, denying everything that goes beyond them, or at least declaring it "unknowable" and refusing to take it into account, which comes to the same thing in practice. This negation existed for a long time as a fact before there was any question of erecting it into a systematic theory under names such as "positivism" and "agnosticism," for it may truly be said to be the real starting point of all modern science. (Guénon 1942, 64–66)

This view of modern science gained further clarity in the lucid prose of Frithjof Schuon. "Modern science, which is rationalist as to its subject and materialist as to its object," he wrote, "can describe our situation physically and approximately, but it can tell us nothing about our extra-spatial situation in the total and real Universe" (Schuon 1965, 111). This "total and real Universe" is seen as beyond the reach of modern science, which is sometimes described as "profane science" to distinguish it from sacred science. "Profane science, in seeking to pierce to its depths the mystery of the things that contain—space, time, matter, energy—forgets the mystery of the things that are contained: it tries to explain the quintessential properties of our bodies and the intimate functioning of our souls, but it does not know what intelligence and existence are; consequently, seeing what its "principles" are, it cannot be otherwise than ignorant of what man is" (Schuon 1965, 111).

The most representative voice of the traditionalist discourse on modern science is that of Seyyed Hossein Nasr. Beginning with *An Introduction to Islamic Cosmological Doctrines* (1964), Nasr's scholarly writings have explored various aspects of Islam's relationship with science over the last forty-two years. These works are part of a corpus of writings that addresses almost all aspects of Islam and its civilization—from the Islamic concept of Ultimate Reality to sacred art and architecture. His works on science have explored the essential features of traditional sciences of nature as well as issues concerning modern science. Nasr's unique position in the Islam and science discourse stems from his thorough training and understanding of modern Western science and traditional Islamic *hikmah* (Wisdom). Ironically, it was during his years at MIT and Harvard that he developed a deep yearning for what was not offered at these prestigious institutions—a Wisdom which could only be learned through an oral tradition. Therefore, soon after his return to his native Iran in 1958, he sought traditional masters so that he could learn wisdom "at their feet" (Nasr 2001b, 41). By that time he had already spent years studying Islam and Western philosophy and had made contact with the great expositors of traditional doctrines such as Schuon and Burckhardt. But it was the period between 1958 and 1979 that proved to be the most important time for his writings on science. His training in the history and philosophy of modern science and the inner resources gathered from traditional wisdom are a unique combination, making him the chief expositor of a clear and insightful Islamic perspective on science. His critique of modern science identifies five main traits of modern science, as Kalin has summed up his position: (i) the secular view of the universe that sees no traces of the Divine in the natural order; (ii) mechanization of the world-picture upon the model of machines and clocks; (iii) rationalism and empiricism; (iv) the legacy of Cartesian dualism that presupposes a complete separation between *res cogitans* and *res extensa*, that is, between the knowing subject and the object to be known; and (v) exploitation of nature as a source of power and domination (Kalin 2001, 453).

Further explaining his position on the "religious view of the cosmos," Nasr rejects the

> external understanding of religion prevalent today as a result of which this phrase means only the acceptance of God having created the world and the world finally returning to God. These truths are of course basic for understanding "the religious view of the cosmos," but they do not include all that this phrase implies. Rather, by "religion" in the term "religious view" here is meant religion in its vastest sense as tradition which includes not only a metaphysics dealing with the nature of the Supreme Reality or Source, but also cosmological sciences which see all that exists in the cosmos as manifestations of that Source, the cosmological sciences themselves being applications of metaphysical principles to the cosmic domain. The religious view of the cosmos relates not only the beginning and end of things in the external sense to God, but also studies all phenomena as signs and symbols of higher levels of reality leading finally to the Supreme Reality and all causes as being related ultimately to the Supreme Cause. (Nasr 2001c, 464)

The traditionalists have produced critiques of modern science in general as well as of its specific theories, in particular the theory of evolution. One of the severest such critiques is to be found in Martin Lings's *The Eleventh Hour* (1988, 15–44). Several logical, scientific, spiritual, and historical arguments are presented by Lings to challenge evolutionism. Lings likened the theory of evolution and that of progress to "two cards that are placed leaning one against the other at the 'foundation' of a card house. If they did not support each other, both would fall flat, and the whole edifice, that is, the outlook that dominates the modern world, would collapse" (Lings 1988, 20). He argued that

> Every process of development known to modern science is subject to a waxing and waning analogous to the phases of man's life. Even civilizations, as history can testify, have their dawn, their noon, their late afternoon, and their twilight. If the evolutionist outlook were genuinely 'scientist', in the modern sense, it would be assumed that the evolution of the human race was a phase of waxing that would necessarily be followed by the complementary waning phase of devolution; and the question of whether or not man was already on the downward phase would be a major feature of all evolutionist literature. The fact that the question is never put, and that if evolutionists could be made to face up to it most of them would drop their theory as one drops a hot coal, does not say much for their objectivity. (Lings 1988, 24)

Whitall N. Perry, another traditionalist, wrote a book on evolution, *The Widening Breach: Evolutionism in the Mirror of Cosmology*, and challenges it from a cosmological standpoint. This refutation states that evolutionism suffers from a *missing link* and that there exists "no prerogative, cosmic principle or law by which this inanimate and subjectless—hence limited—pristine stuff could from its inception maintain over measureless time a perfect self-containment." The author asks:

> The point of all this is to ask simply, why should the pair subject-object alone, on the plane of manifested existence not be a 'pair', but be free from the 'tyranny' of interdependence or linkage to which all the other listed and unlisted terms without exception are subjected? (Perry 1995, 3)

Using a wide range of traditional sources, Perry attempts to place the subject/object polarity in its proper frame of reference. He affirms the primordial truth that the Being of all beings is but one Being and that polarities appear only at the manifest plane of reality. This subject/object relationship is essentially the linchpin for the whole argument against evolutionism, for there can be no object without a subject. Evolutionists may claim that one pole of a duality can exist in the total and unqualified nonexistence of its corollary or counterpart, but such claims cannot be valid for the simple reason that in the whole of the manifest universe not a single example can be found to support this claim. On the other hand, the manifest universe is full of subject/object relationships that are expressed in numerous phenomena—the regularity with which the heavenly objects move, the unerring functioning of all the laws of matter according to their properties, and the interplay of a wide range of dualities to produce logical results in the phenomenal world.

In conclusion, let us mention a last example of the metaphysical discourse on science. The writings of Syed Naquib al-Attas stand apart from the traditionalist school, but there are several common features as well. His writings on the relationship between Islam and science can best be understood within the integrated system of thought he developed on the basis of the application of traditional Islamic philosophy (*hikmah*) to the contemporary situation. Examining science from the metaphysical perspective of Islam means a construction that takes into consideration the authority of revelation, sound traditions of the Prophet, and intuitive faculties granted humanity by the Creator. One key aspect of al-Attas's views on modern science is the epistemological considerations he brings to the discourse. He observes that Islamic metaphysics and modern science are based on two divergent foundations with regard to their respective positions concerning the sources and methods of knowledge. "It is implicit in al-Attas's conception of science as 'definition of reality' that 'science' is to be understood in the wide sense of the term as any objective systematic inquiry, including the intellectual, psychological, natural, social and historical disciplines" (Setia 2003, 172). In his opinion, modern science and philosophy suffer from a myopia that limits our understanding of reality. "God is not a myth, an image, a symbol, that keeps changing with the times," he wrote in his *Islam and the Philosophy of Science*:

> He is Reality itself. Belief has cognitive content; and one of the main points of divergence between true religion and secular philosophy and science is the way in which the sources and methods of knowledge are understood. (al-Attas 1989, 3)

Al-Attas's critique of modern science considers the denial of the reality and existence of God—an implied component of modern science—as the key source of all problems. Modern science conceives the existence of things in terms of their coming into being from other things, as a progression, a development or evolution. This perception of the world construes it as a self-subsistent system evolving according to its own laws.

> The denial of the reality and existence of God is already implied in this philosophy. Its methods are chiefly philosophic rationalism... rationalism, both the philosophic and the secular kind, and empiricism tends to deny authority and intuition as legitimate sources and methods of knowledge. Not that they deny the *existence* of authority and of intuition, but that they reduce authority and intuition to reason and experience. (al-Attas 1989, 6)

The denial or reduction of "the reality and existence of God" is recognized by a large number of Muslim scholars to be the main problem as far as Islam and modern science are concerned. It is not that individual scientists practicing modern science are conceived as not having faith in God; rather, the issue here is the foundational structure of modern science, which leaves out the Creator. All other issues are seen as following this one initial divergence.

ISLAM AND THE BRAVE NEW WORLD OF BIOGENETICS

Perhaps nothing makes the need for a thorough, comprehensive, and creative interaction between religions and modern science as apparent and urgent as certain recent developments in biogenetics and reproductive technologies. After all, we can now have a child come into this world from the womb of a mother where the would-be baby was implanted as an embryo created from the ova of a donor (paid to produce a number of ova following the use of hyper-ovulation medication) and sperm obtained from a sperm bank. The mother of the baby in this actual case was, however, only providing gestational services to a couple who could not have children for medical reasons. Who would be considered the child's real parents? What rights would the donor have over the child? What if one day the donor of the sperm claimed his rights over the child? What would this child inherit? From whom? What if the couple who had paid for the services of the surrogate mother and all other expenses divorced after the child was born—who would have legal rights over the child? What if both of them wanted to keep the child? What if neither wanted the child anymore? What if the sperm donor suddenly changed his mind and decided that the use of his sperm was not in full compliance with the conditions he had set at the time of donation? What is a specific religion's position on these issues?

Let us consider a few more details of this case. As is usual, in vitro fertilizations (IVF) of sperm and ova produced not one but seven embryos. Three of these were placed in the uterus of the woman who offered gestational services for money; pregnancy resulted with a single fetus. The remaining embryos were frozen for possible future use. Two years after the birth of the child, the couple sought and obtained divorce. The baby boy lived with the "mother," who now wanted to have a sibling for the baby boy from the frozen embryos that were preserved in the fertility clinic. The "father," however, objected to the use of these embryos. The agreement signed by the couple at the time of IVF required the consent of both of them for any future use of the frozen embryos. The "mother" filed a lawsuit on the grounds that "her" baby boy was being deprived of his siblings from the gestation of the frozen embryos.

On what grounds can a court of law decide who has the right to what? On what basis can a given religious tradition decide various questions arising from this situation? Religious texts and traditions need to be interpreted for obtaining answers for this case. Who can interpret them if recognized religious scholars do not have adequate scientific understandings of the case?

Such questions had never existed prior to the development of assisted reproductive technologies (ART) now available. Similar questions have been raised by advances in stem cell research, cloning, and biogenetics. Changes in the genetic structure of plants have become more or less a common feature of modern agriculture and animal husbandry, but scientific research has rapidly moved on to the application of genetic engineering to humans and this is raising fundamental questions which before were not even conceivable. Transgenic genetic engineering—that is, the formation of one creature including genetic material from two different species—is no more merely an imaginative leap. These rapid advances have forced all religious traditions to formulate answers to the multiple religious and ethical questions springing forth from these developments in science and technology. The enormity of the issues involved can be judged from specific criminal cases now in courts. These range from theft of recently buried cadavers to the unethical use of placentas and umbilical cords to produce stem cells for huge profit.

When a sperm and an ovum fuse together, a fertilized egg is produced. This begins to divide, ultimately yielding the full human body. In the very early phases of division, the daughter cells are "totipotent"—each may be capable of starting off as if it were the mother cell, to yield a complete individual. Later generations of cells, which cannot give a total body but can, under special treatment, be directed to produce certain tissues or organs, are called "multipotent" or "stem cells." Stem cells produced from adult blood or placenta and umbilical cord blood are helpful in the treatment of a limited number of diseases, while stem cells recovered from the early embryo hold the promise of curing a wider range of known diseases. This has led to a tendency of destroying a living embryo for harvesting stem cells.

The sequence and timing of research has dictated a certain order to the dilemmas. For instance, IVF techniques have produced thousands of frozen embryos stored in various laboratories around the world. Later it became known that frozen embryos have a high incidence of pathogenic mutations. With advances in stem cell research it became obvious that these embryos can be used to harvest stem cells. Does this not amount to destroying a living embryo?

These dilemmas are not merely ethical and religious in nature but also legal, requiring governments to legislate laws. In the Western world, legislation on these issues has emerged from an institutional structure already in place. These procedures have allowed the participation of scientists, lawmakers, and all kinds of religious and public opinions to play an active role in the process. Whether or not a certain government allows and funds stem cell research has been determined through these existing procedures, and decisions remain open to future modifications. For example, Canada had no laws or guidelines to govern stem cell research or federal funding until March 2002, when the Canadian Institutes of Health Research (CIHR) announced guidelines for human pluripotent stem cell research. These were adopted by the major federal funding agencies; it was agreed that no research with human pluripotent stem cells would be funded without the prior review and approval of the Stem Cell Oversight Committee (SCOC) in conformity with CIHR guidelines. While the CIHR was developing these guidelines, the federal government had its committees working on legislations on assisted human reproduction, including the use of human embryos for research. After due process, Bill C-6 (Act Respecting Assisted Human Reproduction and Related Research) became law in March 2004. The SCOC and CIHR's electronically accessible national registry of human embryonic stem cell lines generated in Canada are to play an important role in Canadian stem cell research. The guidelines, effective June 28, 2006, and related issues are available on the CIHR's website (http://www.cihr-irsc.gc.ca/cgi-bin/print-imprimer.pl).

In 2001, the President of the United States announced federal policy for the funding of stem cell research, which allowed federal funding for research using the sixty existing stem cell lines that have already been derived, but it did not sanction or encourage the destruction of additional human embryos. The rationale was that embryos from which the existing stem cell lines were created had already been destroyed and no longer have the possibility of further development as human beings. It was believed that federal funding of these existing stem cell lines would allow scientists to explore the potential of this research for the lives of millions of people who suffer from life-destroying diseases, without destroying the life of further potential human beings. The creation of a new Presidential Council on Bioethics was also announced. The Council's mandate was "to study the human and moral ramifications of developments in biomedical and behavioral science and technology and to study such issues as embryo and stem cell research, assisted reproduction, cloning, genetic screening, gene therapy, euthanasia, psychoactive drugs, and brain implants" (http://www.whitehouse.gov/news/releases/2006/07/20060719-3.html).

On July 19, 2006, the President of the United States vetoed H.R. 810, the "Stem Cell Research Enhancement Act of 2005," because he felt that it had crossed a certain moral limit and that it would mean that "American taxpayers for the first time in our history would be compelled to fund the deliberate destruction of human embryos. Crossing this line would be a grave mistake and would needlessly encourage a conflict between science and ethics that can only do damage to both and harm our nation as a whole" (http://www.whitehouse.gov/news/releases/2006/07/20060719-3.html).

Likewise, on July 24, 2006, the European Union decided to continue funding under new rules adopted by the 25-nation bloc. These rules prevent human cloning and prohibit destroying embryos.

These examples suffice in pointing out the importance of the role of existing or newly created institutions, committees, and procedures in the emergence of laws and guidelines. These processes are a regular feature of the Western political and social system and there is a remarkable degree of integration of these with the institutions of scientific research, though the two may not have similar positions on various issues. A remarkable feature of most countries in the contemporary Muslim world is the absence of such integrated procedures and mechanisms at the governmental level. This should not be surprising, because these new fields of research have little relevance to the level of science and technology present in the Muslim world, where the greatest issue is the provision of potable water.

This is not to say that the entire Muslim world has ignored the need, as new institutions and consultative bodies have come into existence in some Muslim countries such as Egypt, Iran, and Saudi Arabia. In some cases, Muslim scholars have taken personal initiatives. These initiatives have given birth to a small body of literature pertaining to "Islamic perspectives" on the entire range of issues arising out of advanced scientific and technological research. This development is interesting because, on the one hand, these issues are irrelevant to a large majority of Muslims, and on the other hand a very small segment is participating in advanced biomedical research and is aware of the need for clear Islamic perspectives on these issues.

In seminars and conferences on such issues, one often hears statements like the following: "It is high time for Muslims to come up with ... " These well-meaning statements display the same psychological dilemma that had informed the Islam and science discourse during the nineteenth and twentieth centuries. This "catching up syndrome" has created an insatiable hunger for Western science that continues in the contemporary Islam and science discourse. This desire to produce "Islamic perspectives" on contemporary biomedical research is not concerned with the historical process that has generated these issues in the Western world; it only wishes to be present at the forefront of debates to prove that Islam does not lack a position in this respect, regardless of the relevance of the issues to the great mass of Muslims. A new culture of conferences and seminars has, therefore, come into existence in which Muslim scholars attempt to catch up with the Western dilemmas of modern biomedical research. Regardless of the basic aspects of the situation, an Islamic discourse on these issues has been produced.

Since there are very few qualified scholars who can express opinions on matters requiring religio-legal rulings—rather than moral and ethical opinions—the process has required cooperation between religious scholars and scientists. It should be remembered that legal rulings in Islam can only be issued by a person duly trained in Islamic Law (*Shari'ah*). A jurisconsult (*mufti*) who issues a *fatwa* (a nonbinding legal opinion) needs to understand the entire range of issues in their complexity before passing a *fatwa*. This cooperation has become necessary because the small number of Muslim scientists working in Western laboratories (which might be physically located in Muslim lands) in these areas of advance research are themselves not capable of passing these rulings, because they are a product of an educational system that provides absolutely no training in Islamic law. Their personal opinions are, therefore, of little value as far as the *Shari'ah* is concerned. Thus, in order to circumvent this difficulty, the existing discourse solicits the services of a *mufti*. Those who provide this service often do not understand the enterprise of modern science and, therefore, even if they are told the particular details of specific procedures involved in a certain kind of biomedical research, their approval or disapproval lacks a fuller understanding of the scientific aspects of the issue. Factors such as the direction of scientific research in a given polity, the wide range of economic, social, and political aspects of this research, the complexities arising out of the involvement of multinational drug companies and governmental funding agencies, and issues of patents and rights, all of which are not external to the legal ruling being asked of a *mufti*, seldom come into the purview of the religious scholar or even of scientists. The religious educational institutions that have produced these scholars are still grounded in the tenth century, as it were, and are being asked to express their rulings on issues arising out of scientific research of the twenty-first century.

While these issues are essentially irrelevant to the current realities of the Muslim world, the small body of literature dealing with "Islamic perspectives" on such issues is often termed pioneering work in the field. It attempts to derive its principles on the basis of well-known principles of Islamic *ijtihad*, the process through which one derives ruling on matters for which direct precedent is not available in the two primary sources of Islam, the Qur'an and the Sunnah of the Prophet. These principles and their applications have existed for centuries and were the logical place to go for Muslim scholars interested in searching out answers to such new questions. They had already used these principles in regard to certain other aspects of modern science. For example, when autopsies were first introduced to the Muslim world, it became necessary to decide whether or not they were *halal* (permissible) according to Islamic Law. The use of these principles in numerous new situations arising out of scientific research was, in fact, a common practice during the premodern period. Then, however, Muslim jurists (*fuqaha*) were often part of the scientific community, and even when they were not the science itself was in harmony with the basic principles of Islamic law, and hence there was a symmetrical relationship between the astronomy, medicine, physics, mathematics, and geometry cultivated in the Islamic civilization and the *Shari'ah*, the Islamic Law. This harmony was broken in the post–seventeenth century period and Muslim jurists were forced to derive fresh insights into various issues that had started to emerge with the arrival of modern science and technologies in Muslim world.

With regard to the human body, these issues arose in the context of autopsies and other post-mortem investigations, organ transplantations, and the like. Initially, Muslim scholars considered postmortem investigations illegal on the grounds that they violated the dignity of the dead. They could easily find examples from the Qur'an and *Sunnah* to support their conclusion. But as certain benefits of these investigations became evident to them, their objections disappeared, again on grounds for which they could easily find support in the two primary sources of Islam. Now no one objects to autopsies on legal grounds as and when it is required for forensic investigation or to determine the cause of death.

Over a period of time, a certain pattern seems to have emerged in the responses of Muslim jurists to issues arising out of new scientific and technological developments: most jurists first oppose the practice whereas a minority allows it; with time, the practice becomes prevalent and a de facto acceptance is then granted.

The use of the camera to take photographs of human beings is a classical example. As an example of the procedure adopted by Muslim scholars to issue legal rulings, let us consider an Egyptian *fatwa* on organ transplantation issued in 1979 by Shaykh Jad al-Haqq, the *mufti* of Egypt at the time. The first point to understand in this regard is that an issue is only open to juristic discretion (*ijtihad*) ab initio if it does not have any precedent in law. The second point to note is the application of the well-established principle of "consideration of dominant public interest" (*ri'aya masalih al-rajiha*) or common good. The third point of importance in the approach of this *fatwa* is the application of the doctrine of that in the absence of any prohibition specified by Islamic Law, the doctrine of "original permissibility" (*ibaha asliyya*) applies to all situations, and organ transplantation is no exception. "The *fatwa* argues that organs severed from a body are not defiled and advances the view that a believer's body cannot be permanently defiled whether living or dead. Faced with a life-threatening danger it was even permissible to eat human flesh" (Moosa 2002, 336).

Like many *fatwas* dealing with a new situation, the Egyptian *fatwa* discussed analogous precedents to establish grounds for comparing the organ transplantation with other situations. "The *fatwa* argued that by way of *argumentum a fortiori* there was an even greater reason to approve the permissibility of organ transplantation" (Moosa 2002, 337). Here is the relevant part of the *fatwa*:

It is permissible to cut the abdomen of a person and remove an organ or part of it, in order to transplant it to another living body, given the physician's view based on dominant probability that the recipient (donee) will benefit from the donated organ. [This follows] the jurists' consideration of the preponderant public interest, that 'necessity lifts prohibition' and that a 'greater harm can be offset by a lesser harm' and these are authoritative [principles] derived from the noble Qur'an and the sublime Sunna (tradition of the Prophet). (Moosa 2002, 337–38)

Since this 1979 *fatwa*, the questions have become far more complex owing to advances in medical technology, as we saw in the previous example of a child born to a surrogate mother. These complexities demand much more attention to detail, a greater understanding of the scientific procedures, and a far greater comprehension of the legal, ethical, and moral issues involved. Is a person who has been pronounced "brain-dead" but who is kept alive through a "life-support system" still a living human being from the standpoint of Islamic Law? Can an organ be harvested from his body for transplantation? Sometimes, it becomes impossible to apply a general principle to a specific situation to derive straightforward answers. At other times, no clear-cut answer can be found merely on the basis of external circumstances because, depending on the intention of the person, the same act may produce two opposite results.

To deal with these complex situations, certain consultative bodies have been formed. These include the Academy of Islamic Jurisprudence (AJC) established by the Organization of Islamic Conference (OIC), the Islamic Organization for Medical Sciences, the Research Council on Contemporary Issues established in Pakistan by certain Muslim scholars, and various committees and bodies formed by certain Muslim governments under their Ministries of Religious Affairs. These organizations have engaged a number of prominent Islamic scholars and scientists in deliberations about various aspects of new issues arising out of biomedical research. Their "cutting edge" efforts, however, remain peripheral to the larger concerns of the contemporary Muslim world, where most of the issues arising out this research are still like an alien sound coming from distant places.

This irrelevance is notable and so is a peculiar pattern that emerges from this irrelevance. This pattern produces "news"—often international news—which informs the world that such and such Muslim country has now joined the ranks of nations where scientists conduct research on some frontier of biogenetic or reproductive technologies, or stem cell research. In most cases, the small print of such news items is an implantation of a certain research being conducted in the West, often with a complex agenda behind it. At the heart of such efforts are either zealous wealthy patrons who wish to see their country "join the ranks," an international agency, or a corporation hoping to have access to a cheap and unhindered source of human organs, placentas, or blood. This pattern is especially noticeable in research that requires relatively smaller investment in instruments. A certain private center is created in the name of scientific research, a local person trained in the West becomes the director of the center, and soon a full-blown controversy arises in the country, more or less on the pattern of Western controversies. An example of this pattern can be found in the case of an in vitro fertilization center established in 2003 in Cairo. This news appeared in many Western magazines including *The Christian Science Monitor* (June 22, 2005). The center was set up to conduct stem cell research using stem cells from umbilical cord blood, with the hope of using surplus early embryos from consenting couples who no longer need them for future in vitro fertilization. The news item then goes immediately to the sensational aspect of the center: this research "could spark the same kind of ethical debate in Egypt that's now raging in the United States, and the prospect provides a window into the Muslim world's divided views about the issue." From here, it is merely a step to a host of political, social, and cultural issues that can be tagged to such news items. "Some Muslims in Egypt, a deeply conservative and religious country, are open to allowing embryonic stem-cell research," *The Christian Science Monitor* tells its readers, "saying the embryo does not have a soul until later stages in its development. But others agree with Coptic Orthodox and Catholic clergy, who say it is immoral, even infanticide, to destroy embryos at any stage to harvest stem cells."

Standard formula for such news is to mention the perspective of "hardliners," "moderates," and "moderns," attach a host of political statements to the news, and spice it with some statements about Islamic law that present Islam as opposing advanced scientific research. This recipe is used in print and electronic media and it often spills over into academic writings. In the end, what we have is no more than a distorted view of reality in which some Islamic scholars are shown to hold favorable views toward research on, say, embryonic stem cells on the basis of the *Shari'ah*, while others oppose it, again on the basis of the *Shari'ah*. This muddled zone of the Islam and science discourse is a product of its dislocation, its sheer lack of foundation. Often the writer will inform the readers that, unlike in Catholicism, there is no institution or individual who can speak for Islam. This would supposedly render all opinions equally valid for consideration, thus leaving the reader in utter confusion.

This caricature of the Islam and science discourse is a recent invention. It has found popularity for it satisfies certain psychological needs of Muslims as well as Western journalism. It makes Muslims feel at par with "advanced nations" in science; for various Western journalists and media, such news items make economic and political sense. What remains unsaid in all of this is the utter irrelevance of the issue to the vast majority of Muslims, the absence of any real scientific infrastructure in the Muslim world, and the desperate attempts of some Muslim countries to "catch up" with the West in science by some magic stem cell that would remove structural inadequacies and deficiencies of human resources generated over three centuries.

Today, Egypt, Iran, Turkey, Saudi Arabia, Pakistan, and Malaysia conduct stem cell research. Iranian scientists had developed human embryonic stem cell lines as early as 2003. *Fatwas* exist that consider both the therapeutic cloning of embryos as well as embryonic stem cell research lawful. But in countries where the monthly income of most people is from $10 to 50, the danger of selling organs, embryos, and other "spare parts" is high, especially because legal safeguards are practically absent.

Notwithstanding these broader issues, a number of works have come into existence that explore Islam's position on various branches of modern biomedical sciences. Numerous conferences and seminars on these issues have been held during the last decade. They have helped to sharpen certain questions and answers. A representative example of Muslim opinions on these issues can be found in the proceedings of a series of conferences organized by the Islamic Organization for Medical Sciences, Kuwait (http://www.islamset.com/ioms/main.html).

IN CONCLUSION

Islamic perspectives on modern science are intertwined with a host of other political, social, and economic issues. We have examined some of these contributing factors that have shaped the contemporary Islam and science discourse. Two important factors stand out from the rest: Islam's encounter with modernity and a deep-seated, almost insatiable, hunger for modern science in the Muslim psyche. In the final analysis, modern science is a Western enterprise, with deep roots in the Western civilization. Notwithstanding the claims of the universality of modern science, this enterprise cannot be dissociated from the broader cultural matrix from which it emerged. Seen in this perspective, Muslim scholars have an enormous unfinished agenda at hand: to address what is to be done with a science (and the technologies produced by its application) that has obliterated all other means of investigating nature and that has become the most important enterprise in human history in terms of its effect on the way we now live.

In the post-1950 era a new awareness among a small minority of Muslim scholars has produced penetrating critiques of modern science as well as Muslim attitudes toward it, but this strand of discourse remains peripheral to the official attitudes of Muslim states as well as to the general Muslim response to modern science, both of which see modern science from the point of view of its utility, with an almost total disregard to the wider spiritual, cultural, and social implications of importing modern science and technology. The failure of Muslim states to jumpstart a scientific research in their own countries and the enormous social and cultural dislocations modern technologies have produced in many Muslim countries have not led to any reconsideration of attitudes toward Western science and technologies. For those Muslim states that can afford to pay for the most advanced technology that appears on the horizon, there is never a question of considering its impact on society. This headlong plunge into the ethos of the twenty-first century has contributed to a cultural schizophrenia in these countries, where a large majority of the population remains alien to modernity in its attitudes while a small minority pushes these societies into a fast-track process of modernization through the importation of science and technology.

It must be clear by now that it is not Islam's attitude toward science that is under consideration in these cases; it is mostly the psychological complexes of Muslims that have generated this sound and fury in the discourse. As far as Islam is concerned, modern science and technologies based on it cannot be seen as neutral. Brought into the matrix of Islamic metaphysical and moral and ethical principles, modern science and technologies do not remain value-free. As a system of thought as well as one of the most important factors in shaping the way we live, modern science and technologies have to be seen in terms of their impact on society. This impact, let us note, is not merely in terms of certain ethical issues arising out biogenetics, but are of a much broader nature. A small gadget like the cell phone which fits into one's pocket can be as disruptive to a way of life as a complicated procedure that transplants a fetus into the womb of a surrogate mother.

A critique of modern science is often considered an "anti-science" attitude, a sign of conservatism, even fundamentalism. Seen in the context of the violent events that have marked the beginning of the twenty-first century, Islam and science discourse is likely to become even more complicated. Yet, almost two centuries of the clamor of reformers asking Muslims to jumpstart the production of science in their societies has clearly shown that this is not possible, no matter how much science and technology is imported. What is needed is a major intellectual revolution in the Muslim world that would recover the lost tradition of scholarship rooted in Islam's own primary sources. This would lead to the emergence of a new movement helping Muslims to appropriate modern science and technologies like the movement that digested an enormously large amount of scientific and philosophical thought that entered the Islamic tradition during the three centuries of the earlier translation movement. Only such a recasting of modern scientific knowledge has the hope of germinating the seeds of a scientific thinking in the Muslim mind that is not laden with scientism. Only such a revolutionary change in thinking can liberate the Islam and science discourse from its colonized bondage and produce genuine Islamic reflections on the enterprise of modern science—an enterprise that looms large in all spheres of contemporary life and society.

Primary Sources

The following material selected from primary sources has been chosen to provide insights into various aspects of the Islam and science relationship that existed in the Islamic intellectual tradition before the rise of modern science. Many other works of importance, which could not be included here for reasons of space, are listed in the annotated bibliography.

—1—
Ya'qub ibn Ishaq al-Kindi, *On First Philosophy* (*fi al-Falsafah al-Ula*), Translated by Alfred L. Ivry, Albany: State University of New York Press, 1974, pp. 70–75.

Abu Yusuf Ya'qub ibn Ishaq al-Kindi was born in Kufa toward the end of the eighth century or the beginning of the ninth to the governor of Kufa. His family was originally from the southern part of Arabia. His tribe, Kindah, was distinguished by the presence of a Companion of the Prophet. By the time al-Kindi completed his early studies, Baghdad had become the intellectual capital of the world. Thus it was natural for him to go to Baghdad, where he enjoyed the patronage of the caliphs al-Ma'mun and al-Mu'tasim, the latter appointed him his son's tutor. Al-Kindi remained in high positions throughout his life except for the final years when court intrigues led to misunderstanding between him and caliph Mutawakkil.

This led to the confiscation of his large library and a beating—both incidents are sometimes taken as indication of persecution, even as proofs for "Islam against science and philosophy" doctrine. Historical data, however, clearly indicates that competing social, ethnic, and political interests were behind the episode. When he died around 870, al-Kindi's fame spread throughout the Muslim world and he was honored with the title of "the philosopher of the Arabs." Although influenced by Aristotle, al-Kindi is an independent thinker who maintains several important and significant deviations from Aristotelian philosophy. He rejects the idea of the eternal universe, for instance. His concept of causality is also different from Aristotle, because he points to a fifth kind of causality. The following selection from his *On the First Philosophy* presents his views on "motion."

~~~~~~~~

Motion is the motion of a body only:

If there is a body, there is motion, and otherwise there would not be motion. Motion is some change: the change of place (either) of the parts of a body and its center, or of all the parts of the body only, is local motion; the change of place, to which the body is brought by its limits, either in nearness to or farness from its center, is increase and decrease; the change only of its predicate qualities is alteration; and the change of its substance is generation and corruption. Every change is a counting of the number of the duration of the body, all change belonging to that which is temporal.

If, therefore, there is motion, there is of necessity a body, while if there is a body, then there must of necessity either be motion or not be motion.

If there is a body and there was no motion, then either there would be no motion at all, or it would not be, though it would be possible for it to be. If there were no motion at all, then motion would not be an existent. However, since body exists, motion is an existent, and this is an impossible contradiction and it is not possible for there to be no motion at all, if a body exists. If furthermore, when there is an existing body, it is possible that there is existing motion, then motion necessarily exists in some bodies, for that which is possible is that which exists in some possessors of its substance; as the (art of) writing which may be affirmed as a possibility for Muhammad, though it is not in him in actuality, since it does exist in some human substance, i.e., in another man. Motion, therefore, necessarily exists in some bodies, and exists in the simple body, existing necessarily in the simple body; accordingly body exists and motion exists.

Now it has been said that there may not be motion when a body exists. Accordingly, there will be motion when body exists, and there will not be motion when body exists, and this is an absurdity and an impossible contradiction, and it is not possible for there to be body and not motion; thus, when there is a body there is motion necessarily.

It is sometimes assumed that it is possible for the body of the universe to have been at rest originally, having the possibility to move, and then to have moved. This opinion, however, is false of necessity: for if the body of the universe was at rest originally and then moved, then (either) the body of the universe would have to be a generation from nothing or eternal.

If it is a generation from nothing, the coming to be of being from nothing being generation, then its becoming is motion in accordance with our previous classification of motion, (viz.) that generation is one of the species of motion. If, then, body is not prior (to motion, motion) is (of) its essence and therefore the generation of a body can never precede motion. It was said, however, to have been originally without motion: Thus it was, and no motion existed, and it was not, and no motion existed, and this is an impossible contradiction and it is impossible, if a body is a generation from nothing, for it to be prior to motion.

If, on the other hand, the body (of the universe) is eternal, having rested and then moved, it having had the possibility to move, then the body of the universe, which is eternal, will have moved from actual rest to actual movement, whereas that which is eternal does not move, as we have explained previously. The body of the universe is then moving and not moving, and this is an impossible contradiction and it is not possible for the body of the universe to be eternal, resting in actuality, and then to have moved into movement in actuality.

Motion, therefore, exists in the body of the universe, which, accordingly, is never prior to motion. Thus if there is motion there is, necessarily, a body, while if there is a body there is, necessarily, motion.

It has been explained previously that time is not prior to motion; nor, of necessity, is time prior to body, since there is no time other than through motion, and since there is no body unless there is motion and no motion unless there is body. Nor does body exist without duration, since duration is that in which its being is, i.e., that in which there is that which it is; and there is no duration of body unless there is motion, since body always occurs with motion, as has been explained. The duration of the body, which is always a concomitant of the body, is counted by the motion of the body, which is (also) always a concomitant of the body. Body, therefore, is never prior to time; and thus body, motion and time are never prior to one another.

It has, in accordance with this, already been explained that it is impossible for time to have infinity, since it is impossible for quantity or something which has quantity to have infinity in actuality. All time is therefore finite in actuality, and since body is not prior to time, it is not possible for the body of the universe, due to its being, to have infinity. The being of the body of the universe is thus necessarily finite, and it is impossible for the body of the universe to be eternal.

We shall, moreover, show this by means of another account—after it has been explained by what we have say—which shall add to the skill of the investigators of this approach in their penetration (of it). We therefore say:

Composition and combination are part of change, for they are a joining and organizing of things. A body is a long, wide, deep substance, i.e., it possesses three dimensions. It is composed of the substance which is its genus, and of the long, wide and deep which is its specific difference; and it is that which is composed of matter and form. Composition is the change of a state which itself is not a composition; composition is motion, and if there was no motion, there would not be composition. Body is, therefore, composite, and if there was not motion there would not be body, and body and motion thus are not prior to one another.

Through motion there is time, since motion is change; change is the number of the duration of that which changes, and motion is a counting of the duration of that which changes. Time is a duration counted by motion, and every body has duration, as we said previously, viz., that in which there is being, i.e., that in which there is that which it is. Body is not prior to motion, as we have explained. Nor is body prior to duration, which is counted by motion. Body, motion and time are therefore not prior to one another in being, and they occur simultaneously in being. Thus if time is finite in actuality, then, necessarily, the being of a body is finite in actuality, if composition and harmonious arrangement are a kind of change, though if composition and harmonious arrangement were not a kind of change, this conclusion would not be necessary.

Let us now explain in another way that it is not possible for time to have infinity in actuality, either in the past or future. We say:

Before every temporal segment there is (another) segment, until we reach a temporal segment before which there is no segment, i.e., a segmented duration before which there is no segmented duration. It cannot be otherwise—if it were possible, and after every segment of time there was a segment, infinitely, then we would never reach a given time—for the duration from past infinity to this given time would be equal to the duration from this given time regressing in times to infinity; and if (the duration) from infinity to a definite time was known, then (the duration) from this known time to temporal infinity would be known, and then the infinite is finite, and this is an impossible contradiction.

Furthermore, if a definite time cannot be reached until a time before it is reached, nor that before it until a time before it is reached, and so to infinity; and the infinite can neither be traversed nor brought to an end; then the temporally infinite can never be traversed so as to reach a definite time. However its termination at a definite time exists, and time is not an infinite segment, but rather is finite necessarily, and therefore the duration of body is not infinite, and it is not possible for body to be without duration. Thus the being of a body does not have infinity; the being of a body is, rather, finite, and it is impossible for body to be eternal.

It is (also) not possible for future time to have infinity in actuality: for if it is impossible for (the duration from) past time to a definite time to have infinity, as we have said previously; and times are consecutive, one time after another time, then whenever a time is added to a finite, definite time, the sum of the definite time and its addition is definite. If, however, the sum was not definite, then something quantitatively definite would have been added to something (else) quantitatively definite, with something quantitatively infinite assembled by them.

Time is a continuous quantity, i.e., it has a division common to its past and future. Its common division is the present, which is the last limit of past time and the first limit of future time. Every definite time has two limits: a first limit and last limit. If two definite times are continuous through one limit common to them both, then the remaining limit of each one of them is definite and knowable. It has, however, been said that the sum of the two times will be indefinite; it will then be both not limited by any termini and limited by termini, and this is an impossible contradiction. It is thus impossible, if a definite time is added to a definite time, for the sum to be indefinite; and whenever a definite time is added to a definite time, all of it is definitely limited, to its last (segment). It is, therefore, impossible for future time to have infinity in actuality.

—2—
Ibn Sina–al-Biruni Correspondence, *al-As'ilah wa'l-Ajwibah*, Translated by Rafik Berjak and Muzaffar Iqbal, *Islam and Science*, Vol. 1, Nos. 1 and 2, 2003, pp. 91–98 and 253–60

Writing from Khwarizm, the modern Khiva and ancient Chorasmia, Abu Rayyhan Muhammad b. Ahmad al-Biruni (973–1050) posed eighteen questions to Abu Ali al-Hasayn b. 'Abd Allah ibn Sina (980–1037). Ten questions were related to various concepts and ideas in Aristotle's *De Caelo*. Ibn Sina responded, answering each question one by one in his characteristic manner. Not satisfied by some of the answers, al-Biruni wrote back, commenting on the first eight answers from the first set and on the seven from the second. This time, the response came from Abu Said Ahmad ibn Ali al-Masumi, whose honorific title, Faqih, is indicative of his high status among the students of Ibn Sina. He wrote on behalf of his teacher, who was the most representative scholar of Islamic Peripatetic tradition. This encounter was rigorous and indicative of many important aspects of Islamic philosophical and scientific traditions. It also shows how Muslim scholars and scientists worked out certain basic concepts in natural sciences.

~~~~~~~~

In the name of Allah the Most Merciful the Most Compassionate.

1. **The Grand Master, Abu Ali al-Hussein Abu Abdullah Ibn Sina**—may Allah grant him mercy—said, All Praise is for Allah, the Sustainer of the worlds, He suffices and He is the best Disposer of affairs, the Granter of victory, the Supporter. And Allah's blessings be upon our master Muhammad and upon his family and all his companions, and now to begin:

2. This letter is in response to the questions sent to him by Abu Rayhan al-Biruni from Khwarizm. May Allah surround you with all you wish for, and may He grant you all you hope for and bestow on you the happiness in this life, and hereafter, and save you from all you dislike in both lives. You requested—may Allah prolong your safety—a clarification about matters some of which you consider worthy to be traced back to Aristotle, of which he spoke in his book, *al-Sama' wa'l-Alam (De caelo)*, and some of which you have found to be problematic. I began to explain and clarify these briefly and concisely, but some pressing matters inhibited me from elaborating on each topic as it deserves. Further, the sending of the response to you was delayed, awaiting al-Masumi's dispatch of letter to you. Now, I would restate your questions in your own words, and then follow each question with a brief answer.

3. **The first question**: You asked—may Allah keep you happy—why Aristotle asserted that the heavenly bodies have neither levity nor gravity and why did he deny absence of motion from and to the center. We can assume that since the heaven is among the heaviest bodies—and that is an assumption, not a certainty—it does not require a movement to the center because of a universal law that applies to all its parts judged as similar. If every part had a natural movement toward the center, and the parts were all connected, then it would result in a cessation (*wuquf*) [of all motion] at the center. Likewise, we can assume that the heaven is among the lightest of all bodies, this would not necessitate (i) a movement from the center until its parts have separated and (ii) the existence of vacuum outside the heaven. And if the nonexistence of vacuum outside the heaven is an established fact, then the heaven will be a composite body like fire. [And you also say] that the circular movement of the heaven, though possible, might not be natural like the natural movement of the planets to the east [which] is countered by a necessary and forceful movement to the west. If it is said that this movement is not encountered because there is no contradiction between the circular movements and there is no dispute about their directions, then it is just deception and argument for the sake of argument, because it cannot be imagined that one thing has two natural movements, one

to the east and one to the west. And this is nothing but a semantic dispute with agreement on the meaning, because you cannot name the movement toward the west as opposite of the movement to the east. And this is a given; even if we do not agree on the semantics, let us deal with the meaning.

4. **The answer**: May Allah keep you happy, you have saved me the trouble of proving that heaven has neither levity nor gravity, because in your prelude you have accepted that there is no place above the heaven to where it can move, and it cannot, likewise, move below because all its parts are connected. I say it is also not possible for it to move down, nor is there a natural place below it to where it can move, and even if it were separated—and we can make the assumption that it is separated—it would result in the movement of all the elements from their natural positions and this is not permissible, neither by the divine nor by the natural laws. And that would also establish vacuum which is not permissible in the natural laws. Therefore, the heaven does not have a natural position below or above to which it can move in actuality (*bi'l fil*) or in being, neither is it in the realm of possibility (*bi'l-imkan*) or imagination (*bi'l-wahm*) because that would lead to unacceptable impossibilities we have mentioned, I mean the movement of all the elements from their natural positions or the existence of vacuum.

5. There is nothing more absurd than what cannot be proved to exist either by actuality or by possibility or imagination. If we accept this, it follows that heaven does not have a natural position, neither at the top nor at the bottom. But every body has a natural position. And to this, we add a minor term and that is our saying: "heaven is a body", and hence, it will follow from the first kind of syllogism (*shakl*) that heaven has a natural position. And if we could transfer the conclusion to the disjunctive positional syllogism, we could then say: its natural position is above or below or where it is. And if we hypothesize the negation of its being either above or below, we could say: it is neither up nor down; hence the conclusion is: it is where it is.

6. Everything in its natural position is neither dense or light in actuality and since heaven is in its natural position, it is, therefore, neither light nor dense in actuality. The proof of this is that whatever is in its natural position and is light, it will be moved upward because it is light and its natural position is upward but it cannot be said that whatever is light, is in its natural position in actuality because this will contradict what I have just said: it will be "in its natural position" as well as "not in its natural position" at the same time; and that is self-contradictory. And likewise for the dense. Because the dense is what naturally moves downwards and its natural position is down because anything that moves naturally, its movement takes it toward its natural position. And from the first premise, it is clear that the thing in its natural position is not dense in actuality, so when we add the results of the two premises, the sum of this will be that whatever is in its natural position, is neither dense nor light in actuality. And it was established in the second minor term that the heaven is truly in its natural position, therefore, the correct logical conclusion is that the heaven is neither light nor dense in actuality and it is not so potentially (*bi'l-quwwa*) or contingently.

7. The proof of this is that the light and dense in potentia can be so in two situations: (i) It can be so either as a whole, like the parts of the fixed elements in their natural position, so if they were neither dense nor light in actuality, then they are so potentially, for the possibility of their movement by a compulsory motion which can cause them to move from and to their natural position either by an ascending or descending natural movement; and (ii) by considering the parts as opposed to the whole in the fixed elements. These parts are neither light nor dense in their totalities, because if it would move upward, some of the parts would move downward because they are spherical in their shapes and have many dimensions, but indeed, the levity and density are in their parts, so if the heaven is light or heavy potentially, that is in its totality—and we have proved that by nature, the upward or downward movement of the heaven is negated (*maslub*) to its totality, and to prove that we depended on some of your premises. So it was made clear to us that the heaven in its totality is neither light nor dense. And I say that it is neither heavy nor light potentially in its parts because the levity and the density of the heavy and the light parts appear in their natural movement to their natural position. And the parts which are moving to their natural position move in two cases: (i) they might be moving from their natural position by force, [in which case] they would move back to their natural position by nature or (ii) they are being created and moving to their natural position like the fire that emerges from the oil and is moving up. It is not possible for a part of the heaven to move from its natural position by force because that requires an outside mover, a corporeal or non-corporeal mover that is not from itself.

8. The non-corporeal movers, like what the philosophers call nature and the active intellect (*al-aql al-faal*), and the First Cause (*al-illatul ula*), are not supposed to create forced movement (*harakah qasriyyah*); as for nature, it is self-evident, and as for the intellect and the First Cause, their inability [to do so] is left to the Divine knowledge. As for the physical cause, it should be, if possible, one of the [four] elements or composed of them because there is no corporeal body other than these five—the four simple elements and [the fifth being] their combination.

9. And every body that moves by itself and not by accident, moves when it is touched by an active mover. And this has been explained in detail in the first chapter in the book of *Generation and Corruption (Kitab al-kawn wa'l-fasad)*. Thus, it is not possible for a part of the heaven to move without being touched by the mover during its movement toward it either by force (*bi'l-qasr*), or by nature (*bi'l-tab*). The outside mover that moves it by force has to be connected to another mover, which in turn, has to be connected to the first mover of all. And if it was moving by nature, it will be either the non-composite fire or a combination in which the fire-parts are dominant. The non-composite fire does not affect the heaven because it engulfs it from all sides and the impact of bodies on bodies is by touch and there is no part in the heaven which is more passive than the other, unless one of the parts is weaker in its nature. However, the weakness of the substance does not come from itself but through an outside factor.

10. Thus, the question now returns to the beginning, to that of a compound mover in which the fire-part is dominant. It will not have impact until it reaches the sphere of the heaven and when it reaches the airy zone, then it will turn into pure fire and burst into a flame as seen in the case of comets. And if it is too slow to reach that transforming stage, it would not touch the heaven, [it may be so] because in it are dense parts, earthly and others, which have gravity. Thus, it is not possible for anything to touch the heaven except pure fire. It is possible for pure or non-pure fire—and the compound is not pure fire—and for the one that is not pure fire it is possible for it to be in the neighborhood of the three elements but it is not possible for it to touch the heaven by nature.

11. As for the other elements, it is not possible for them to touch the heaven in their totality because they do not move in their totality from their natural position, neither in their compound form nor in their parts, thus, they cannot have any impact on the heaven because they are unable to touch it because when they reach the ether (*al-athir*), they will burn and turn into fire and the fire does not touch heaven, as we have proved. But ether changes and disjoins everything that occurs in its [realm] because it is hot in actuality and one of the properties of the hotness in actuality is that it brings together similar genera and separates dissimilar genera—it is the separator of dissimilar and gatherer of similar genera. And when the fire takes over a body that is being affected by it, if it were a compound body made from different parts, the fire will return it to its nature; this shows that [the body] did not change into something that is contrary to its essence by mixing with the affective element. As for the cold, it is not like this. And there is no doubt that the hot is most effective and powerful of all things; and the thing that is in its natural position, strengthens its genus; and the whole is stronger than its parts. So what do you think of something that is hot in its natural position and it is whole, and it allows a part to enter into its sphere and it does not produce any effect [on this part], neither changes it back to its nature, nor separates it, if it were compound?

12. From these premises, it is clear that it is not possible for any part or compound from the elements to reach the heaven. Since they do not reach it, they do not touch it, and if they do not touch it, they do not produce any effect on it. None of the parts or the compounds has any effect on parts of the heaven and if nothing is able to affect it, other than it, from whole or parts, simple or compound bodies, it is not going to be affected and moved potentially by itself. And if we would set aside our premise—and that is our saying, "and it is not possible [for the heaven] to be affected by anything other than by itself", which is true—the result is our saying: "it is not possible that it will be affected and moved by force"; and this is also true. So the heaven is neither light, nor dense potentially, neither as a whole or in its parts. And we have proved that it is not so in actuality. It is neither light, nor dense in general or absolutely. And that is what we wanted to clarify. But you can call the heaven light from the perspective in which people call a floating body, on top of another body, lighter than the latter by nature. So, from this perspective, it is possible that the heaven is the lightest of all things.

13. Now, as to your saying that the circular motion [of the heaven] is natural to it, and your saying, "if it is said that this is not accidentally" et cetera, there is no one among the scholars who has proven the natural circular motion of the heaven, who has ascertained what you have said. I would have explained the reasons, had it not been a separate issue, taking too long [to explain].

14. As for your demonstration that the movement of the stars and the planets is opposite, it is not so. It is only different. Because the opposite movements are opposite in the directions and the ends, and if it was not that the high is opposite of low, then we would not have said that the movement from the center is opposite of the movement to the center; and this has been explained in detail in the fifth chapter of *Kitab al-Sama al-tabii*. As for the directions of the two circular motions and their ends, they are, in our assumption, positional, not natural. Because in nature, there is no end to the circular movement of the heaven, hence it is not opposite; hence the two different circular motions are not opposite and this is what we wanted to clarify.

—3—
Abu Ali al-Husayn ibn Abd Allah Ibn Sina, *The Canon of Medicine (al-Qanun fi'l tibb)*, Adapted by Laleh Bakhtiar from Translation by O. Cameron Gruner and Mazar H. Shah, Chicago: Kazi Publications, 1930 [reprint 1999], pp. 11–14.

The "Prince of Physicians," as Ibn Sina was called by his contemporaries and later generations, from whose *Canon* the following selection has been made, was born in Afshana near Bukhara (in present-day Uzbekistan) in 980 and died in Hamadan in present-day Iran in 1037. *Al-Qanun fi'l tibb*, his magnum opus in medicine, was first translated into Latin by Gerard of Cremona and in spite of enormously large size, it was printed in Latin at least a dozen times before 1501. It was translated into Hebrew by Nathan ha-Me'ati, published in Rome in 1279. The first Arabic edition to be published in Europe is one of the Arabic incunabula (extant copies of books produced in the earliest stages, i.e. before 1501, of printing from movable type), a splendid folio printed by the Typographia Medicca in Rome in 1593 (Sarton 1955, 42). Once translated into Latin, the *Canon* was to quickly gain the status of a classic in medicine. It remained one of the most used texts for the next four centuries. "The fame of Avicenna was so great that medical progress did not shake it and there was still a professional market for the *Canon* during the whole of the seventeenth century"(Sarton 1955, 44).

The following selection from Book I provides insights into Ibn Sina's philosophy of medicine.

∼∼∼∼∼∼∼∼

1.2. Concerning the Subject-Matter of Medicine

§12 Medicine deals with the states of health and disease in the human body. It is a truism of philosophy that a complete knowledge of a thing can only be obtained by elucidating its causes and antecedents, provided, of course, such causes exist. In medicine it is, therefore, necessary that causes of both health and disease should be determined.

§13 Sometimes these causes are obvious to the senses but at other times they may defy direct observation. In such circumstances, causes and antecedents have to be carefully inferred from the signs and symptoms of the disease. Hence, a description of the signs and symptoms of disease is also necessary for our purpose. It is a dictum of the exact sciences that knowledge of a thing is attained only through a knowledge of the causes and origins of the causes, assuming there to be causes and origins. Consequently our knowledge cannot be complete without an understanding both of symptoms and of the principles of being.

1.3. The Causes of Health and Disease

§14 There are four causes—material, efficient, formal, and final. On the subject of health and disease, we have the following:

1.3.1. The Material Causes

§15 The material (*maddi*) cause is the physical body which is the subject of health and disease. This may be immediate as the organs of body together with their vital energies and remote as the humors and remoter than these, the elements which are the basis both for structure and change (or dynamicity). Things which thus provide a basis (for health and disease) get so thoroughly altered and integrated that from an initial diversity there emerges a holistic unity with a specific structure (or the quantitative pattern of organization) and a specific type of temperament (the qualitative pattern).

1.3.2. The Efficient Causes

§16 The efficient (*failiya*) causes are capable of either preventing or inducing change in the human body. They may be external to the person or internal. External causes are: age, sex, occupation, residence, and climate and other agents which effect the human body by contact whether contrary to nature or not. Internal causes are sleep and wakefulness, evacuation of secretions and excretions, the changes at different periods of life in occupation, habits and customs, ethnic group and nationality.

1.3.3. *The Formal Causes*

§17 The formal *(suriyah)* causes are three: temperaments *(mizajat)* (or the pattern of constitution as a whole) and the faculties or drives *(qawa)* which emerge from it and the structure (the quantitative patterns).

1.3.4. *The Final Causes*

§18 The *final (tamamia)* causes are the actions or functions. They can be understood only from a knowledge of both the faculties or drives *(qawa)* and the vital energies (breaths, *arwah*) that are ultimately responsible for them. These will be described presently.

1.4. *Other Factors to Consider*

§19 A knowledge of the above-mentioned causes gives one insight into how the body is maintained in a state of health and how it becomes ill. A full understanding of just how health is conserved or sickness removed depends on understanding the underlying causes of each of these states and of their "instruments." For example, the diet in regard to food, drink, choice of climate, regulations regarding work and rest, the use of medicines, or operative interference. Physicians treat all these points under three headings as will be referred to later: health, sickness, and a state intermediate between the two. But we say that the state which they call intermediate is not really a mean between the other two.

§20 As the aim of medicine is to preserve health and eradicate disease, there are some other factors which deserve consideration: (1) the elements; (2) the temperaments; (3) the humors or body fluids; (4) the tissues and organs-simple and composite; (5) the breaths and their natural, nervous and vital faculty or drives; (6) the functions; (7) the states of the body health, sickness, intermediate conditions; and (8) their causes: food, drink, air, water, localities of residence, exercise, repose, age, sex, occupation, customs, race, evacuation, retention and the external accidents to which the body is exposed from without; (9) the diet in regard to food, drink, medicines; exercises directed to preserving health; and (10) the treatment for each disorder.

§21 With regard to some of these things there is nothing a physician can do, yet he should recognize what they are and what is their essential nature, whether they are really existent or not. For a knowledge of some things, he depends on the doctor of physical science; in the case of other things, knowledge is derived by inference or reasoning. One must presuppose a knowledge of the accepted principles of the respective sciences of origins in order to know whether they are worthy of credence or not; and one makes inferences from the other sciences which are logically antecedent to these. In this manner one proceeds step by step until one reaches the very beginnings of knowledge, namely pure philosophy; to wit, metaphysics.

§22 Things which the medical practitioner should accept without proof and recognize as being true are: (1) the elements and their number; (2) the existence of temperament and its varieties; (3) humors, their number and location; (4) faculty or drives, their number and location; (5) vital forces, their number and location; and (6) the general law that a state cannot exist without a cause and the four causes. Things which have to be inferred and proved by reason are: (1) diseases; (2) their causes; (3) symptoms; (4) treatment; and (5) their appropriate methods of prevention. Some of these matters have to be fully explained by reason in reference to both amount *(miqdar)* and time *(waqt)*.

§23 If a physician like Galen attempted a logical explanation of these hypotheses, he would be discussing the subject not as a medical practitioner, but as a philosopher, and in this way would be like a jurist trying to justify the validity of, say, consensus of opinion. Of course, this he might do, not as a jurist but as a man of knowledge. However, it is not possible either for a medical practitioner, as such, or a jurist in his own capacity, to prove such matters by logic and reason; and if he does so, it will be at his own peril.

§24 The physician must also know how to arrive at conclusions concerning: (1) the causes of illnesses and the individual signs thereof; and (2) the method (most likely to) remove the disorder and so restore health. Wherever they are obscure, he must be able to assign to them their duration, and recognize their phases.

—4—
Abu Bakr Muhammad bin Tufayl, *Hayy ibn Yaqzan*, ed. Léon Gauthier, Beirut, 1936, pp. 70–78, especially translated for *Science and Islam* by Yashab Tur.

Born around 1100 in Wadi Ash, a small town northeast of Granada, Abu Bakr bin Abd al-Malik bin Muhammad bin Muhammad bin Tufail al-Qaisi (known to the Muslim world as Ibn Tufail and to the Latin West as Abubacer) was a man of many gifts. He practiced and taught medicine, and was a philosopher, a mathematician, a poet, and a man of great imagination. He was a teacher of Ibn Rushd, an advisor and court physician first in Granda and later at the court of Prince Abu Sa'd Yusuf, Sultan of the Muwahidin in Morocco, where he died in 1185.

His masterpiece, *Hayy Ibn Yaqzan* (*Living the Son of the Awake*), is an imaginative tale of an infant who is cast ashore upon an equatorial island because his birth has to be concealed. He is suckled by a doe and spends the first fifty years of his life without contact with any human being. His solitary life, which is presented as consisting of seven stages of seven years, is full of reflections on nature amidst continuously emerging new needs which must be met in order to survive. Through his reflections and in the very process of fulfillment of his needs, Hayy is led to such profound concepts as soul and its Creator. This intellectual apprehension of truth is followed by an experiential realization of Reality.

Soon after this leap, Hayy comes into contact with another human being, a spiritually endowed man by the name of Asal who arrives on the island in search of solitude to contemplate the ultimate nature of things. Hayy and Asal find congenial companionship in each other and realize that they have both arrived at the same Truth, each in his own way. Asal tells Hayy about how human society is structured and this produces in Hayy a desire to go to other human beings and show them the real nature of life. A ship arrives in due course of time and they both leave their solitary life for the island where Asal lived and where his friend, Salaman, rules. Hayy preaches to Asal's community, but in time realizes that the truth he has acquired must be acquired by each human being in his or her own personal manner; it cannot be transmitted by preaching. He then departs with Asal to resume a life of contemplation in the solitude of the island where he had lived most of his life.

Ibn Tufail's imaginative tale is an attempt to show that knowledge of the Truth gained through the intellect and reason does not contradict what is apprehended through mystical experience. The tale is based on an earlier work of the same title by Ibn Sina. It is rooted in the Islamic concept of *fitrah*, the innate nature of human beings, and in the Qur'anic message that when left to itself, human intellect is capable of comprehending the ultimate Reality which is none other than One God. What is shown through the experiences of Hayy is, therefore, a confirmation of the process of convergence of sound reasoning, observation, and experiences.

~~~~~~~~

He searched for some common characteristic in all physical objects, animate and inanimate, but could find nothing except their extension in three dimensions. He recognized this as a physical property since all objects were physical. His senses did not find any object which had just this and no other characteristic. On further examining this notion by asking himself [pertinent questions] such as whether or not another principle existed besides extension, he realized that there must be another factor, besides extension, to which extension is attached. For mere extension could no more subsist by itself than the extended object could exist without extension.

Hayy tried out this idea on several objects which had form. When he examined clay, he discovered that if he molded it into some shape, for example into a ball, it had length, width, and depth in a certain proportion; if he then took this ball and formed a cube or an egg-shaped object, its length, width, and depth still existed, but now they took on different proportions, though it was still the same clay. No matter what the proportion of length, width, and depth, the object could not be divested of these properties altogether. Since one proportion could replace another, it became apparent to him that the dimensions were a factor in their own right, distinct from the clay itself. But the fact that the clay was never totally devoid of dimensions made it clear to him that they were part of its being [as clay].

His experiments led him to believe that all bodies are composed of two things: (i) a thing similar to clay in his experiment on the clay, and (ii) the three extensions (length, width, and depth) of the form into which clay or any other object is formed. This could be a ball, a block, or another figure the clay might have. Thus, he realized that he could not comprehend physical things at all unless he conceived of them as compounded of these two factors, neither of which could subsist without the other.

He deduced that there is a variable factor—the form of the bodies. This can have many different faces, and three properties of extension (length, width, and depth) correspond to this form. The other factor, which remains constant like the clay of the example, corresponds to materiality in all other bodies. In philosophy the factor analogous to the clay is called *hyle*, or matter. This is pure matter, devoid of forms.

Now that his thinking had achieved a certain level of sophistication and he could use the faculties of his mind, he felt alien and alone because the sensory world had now receded to some extent. This produced a longing for the familiar world of the senses. He disliked the notion of the unqualified body—a thing he could neither possess nor hold. He reverted to the four simple objects he had already examined.

First he reexamined water. He found that when left to itself, in its own natural form, it was cold and moved downward; but if warmed by fire or the heat of the sun, its coldness departed, only its tendency to fall remained as its property. If it were heated vigorously, this second property also disappeared; in fact, now [water] gained the [opposite] tendency: it rose upward. In this way, both primary properties of water were changed. He concluded that once the two original properties were gone, the form of water must have left this body, since it now exhibited behavior characteristic of some other form. Hence, a new form, not previously present, must have come into existence, giving rise to behavior unlike that it had shown under its original form.

Hayy now knew for sure that all that comes into existence must have a cause. A vague and diffused notion of cause of forms now appeared to him. He went over the forms he had known before and realized that all of these forms had come into existence due to some cause. Examining the essence of each form, he realized that the essence of each form was nothing but the potential of a body to produce actions. The object in which a form would inhere, he thus realized, was a body's propensity for an action. Water has a propensity to rise when heated; this propensity is contained in its form. The capacity of an object to inhere certain actions—and not others—resides in its form.

Hayy concluded that his observation would be true of all forms. Actions emerging from forms did not arise in them; all actions attributed to them were actually brought about through them by another Being. This idea to which Hayy had now achieved certitude is the same as the one found in the Prophet's words: "I am the ears He hears by and the sight He sees with. It is also mentioned in the perfect Revelation: *It was not you but God who killed them; and when you shot, it was not you who shot, but God*. (Q. al-Anfa'al: 17)

Having achieved this general notion of cause, Hayy developed an intense longing to know more about it. He had not yet left the world of senses, hence he first tried to find some ultimate Cause in the sensory world. He did not yet know if this ultimate Cause of all things was one or many. He reexamined all the objects he had examined before and decided that they had all come into existence at some point and then decayed either totally or in part. Water and earth, for example, are destroyed in part by fire. Air changes into snow when it is very cold and snow into water when it warms. Nothing around him was exempt from change, and no change existed without a specific cause.

Hayy now left behind all things he had examined and turned his mind to the heavenly bodies. He had now lived four seven-year cycles and was a man of twenty-eight years.

At the outset, he knew that the heavens and all the stars were bodies, because without exception they were extended in three dimensions, and whatever extended in three dimensions is a body. Therefore they were all bodies. The question he now pondered on was whether they extended infinitely in all directions or were they finite—bounded at some point beyond which there was no extension. This problem perplexed him a great deal, but ultimately his inborn intelligence led him to conclude that an infinite body is a propensity, which can neither exist nor be conceived. This conclusion was confirmed in his mind by a number of factors which he considered through a reflective approach.

"This heavenly body I see," he said to himself, "is without doubt finite on the side which I see; my eyes and all other senses tell me this. I cannot doubt this. As to the side I cannot see, it will either be finite or infinite. But it cannot be infinite because it is impossible for anything to extend forever. If I were to start two lines from the finite side of the body and let them go through the thickness of the body, if it were infinite, the two lines will go on through the thickness to infinity. Suppose, now I cut off a large segment of one line from the finite end, then reexamine both lines.

There are two possibilities: Either both lines still extend to infinity, or the cut line does not extend as far as the uncut line. If they both still extended to infinity, it would mean that the line with the cut off segment is still of the same length as the other—this is not possible as a part cannot be equal to a whole.

"The other possibility is that the line that has been cut is shorter than the uncut line. This would mean that it is finite. The segment that I had cut off is finite. But if that is finite, the whole must be finite. And if this whole line [which was cut into two segments] is finite, the uncut line must also be finite, because both lines were equal to begin with. This can be applied to all physical things. Thus to assume that any object can be infinite is false and absurd."

This is how he reached the conclusion, through his exceptional intelligence, that heavens must be finite. Hayy now longed to discover the shape of the heavens and their limiting distances. Therefore he watched the sun, the moon, and other stars. He observed that they all rose in the east and set in the west. Those which traveled directly overhead inscribed a great arc while those inclining north or south from his zenith, inscribed a smaller arc. The further they lay from the zenith and the closer to the poles, the smaller the arc they traversed. The smallest orbits in which stars moved were those that we call Ursa Minor and Canopus, two little circles about the North and South Poles respectively. As already mentioned, Hayy's island was on the equator, and therefore all these circles—whether they lay to the north or south—were visible to him.

—5—
Abu al-Waleed Muhammad Ibn Rushd, *Tahafut al-Tahafut (The Incoherence of the Incoherence)*, Translated from the Arabic with Introduction and Notes by Simon van den Bergh, 2 vols. London: Messrs. Luzac & Co., 1954, pp. 311–16.

Abu al-Waleed Ibn Rushd (1126–1198), known to the Latin West as Averroes, was called *the commentator* because of his excellent commentaries on Aristotle. He was born to a distinguished family of jurists and was himself a jurist and physician. He produced works on medicine, jurisprudence, and philosophy. His most important work, *Incoherence of the Incoherence*, from which the following excerpt is taken, was written as an extensive response to the famous work of Abu Hamid al-Ghazali (1058–1111), *Incoherence of the Philosophers*. In much of the secondary literature on

Islam and science written in the West, al-Ghazali's *Incoherence of the Philosophers* is often held as the main culprit for the decline of science in Islamic civilization. It is, therefore, interesting to read Ibn Rushd's response to this work, which first quotes al-Ghazali's arguments and then responds to it. In a way, this can be seen as a debate between two of the greatest minds in Islamic tradition, al-Ghazali and Ibn Rushd. It is interesting to note that al-Ghazali's work tackles twenty issues, out of which only the last four are about the natural sciences. The main charge against al-Ghazali is that he destroyed science in Islamic civilization by destroying causal relations. As can be seen from the following excerpt, al-Ghazali is in fact advocating an occasionalist view. He does so to preserve the Islamic view of miracles. The debate is, therefore, not really on science per se, but on the limits of rational inquiry into meta-scientific matters. In terms of causality, al-Ghazali holds that every time fire burns cotton, the fire itself does not produce the burning effects; they are caused directly by God. It is in God's power to stop fire from producing these habitual effects, if and when He so wishes. This accounts for the presence of miracles.

Ibn Rushd responds by pointing out that a denial of direct causation would destroy the fixed natures. If fire no longer has the causal power of burning, then there is nothing to distinguish it from other things such as water or earth. Consequently, we can no longer differentiate one thing from another in any real sense. This amounts to a destruction of peculiar and distinctive nature of individual substances and hence we can no longer have any real knowledge of the natural world. Thus, the removal of the cause-and-effect relationship leads to the removal of the possibility of knowledge of nature.

~~~~~~~~

About the Natural Sciences

Ghazali says:

The so-called natural sciences are many, and we shall enumerate their parts, in order to make it known that the Holy Law does not ask one to contest and refute them, except in certain points we shall mention. They are divided into principal classes and subdivisions. The principal classes are eight. In the first class are treated the divisibility, movement, and change which affect body in so far as it is body, and the relations and consequences of movement like time, space, and void, and all this is contained in Aristotle's *Physics*. The second treats of the disposition of the parts of the elements of the world, namely heaven and the four elements which are within the sphere of the moon, and their natures and the cause of the disposition of each of them in a definite place; and this is contained in Aristotle's *De coelo*. The third treats of the conditions of generation and corruption, of equivocal generation and of sexual generation, of growth and decay, of transmutations, and how the species are conserved, whereas the individuals perish through the two heavenly movements (westwards and eastwards), and this is contained in *De generatione et corruptione*. The fourth treats of the conditions which are found in the four elements through their mixture, by which there occur meteorological phenomena like clouds and rain and thunder, lightning, the halo round the moon, the rainbow, thunderbolts, winds, and earthquakes. The fifth treats of mineralogy, the sixth of botany. The seventh treats of zoology, which is contained in the book *Historic animalium*. The eighth treats of the soul of animals and the perceptive faculties, and says that the soul of man does not die through the death of his body but that it is a spiritual substance for which annihilation is impossible.

The subdivisions are seven: The first is medicine, whose end is the knowledge of the principles of the human body and its conditions of health and illness, their causes and symptoms, so that illness may be expelled and health preserved. The second, judicial astrology, which conjectures from the aspects and configuration of the stars the conditions which will be found in the world and in the State and the consequences of dates of births and of years. The third is physiognomy, which infers character from the external appearance. The fourth is dream-interpretation, which infers what the soul has witnessed of the world of the occult from dream images, for the imaginative faculty imagines this symbolically. The fifth is the telesmatical art, that is the combination of celestial virtues with some earthly so as to constitute a power which can perform marvelous acts in the earthly world. The sixth is the art of incantation, which is the mixing of the virtues of earthly substances to produce marvelous things from them. The seventh is alchemy, whose aim is to change the properties of minerals so that finally gold and silver are produced by a kind of magic. And there is no need to be opposed to any of these sciences by reason of the Divine Law; we dissent from the philosophers in all these sciences in regard to four points only.

I say:

As to his enumeration of the eight kinds of physical science, this is exact according to the doctrine of Aristotle. But his enumeration of the subdivisions is not correct. Medicine is not one of the natural sciences, but is a practical science which takes its principles from physical science; for physical science is theoretical and medicine is practical, and when we study a problem common to theoretical science and practical we can regard it from two points of view; for instance, in our study of health and illness the student of physics observes health and nature as kinds of natural existents, whereas the physician studies them with the intention of preserving the one, health, and keeping down the other, illness. Neither does judicial astrology belong to physical science; it is only a prognostication of future events, and is of the same type as augury and vaticination. Physiognomy is also of the same kind, except that its object is occult things in the present, not in the future. The interpretation of dreams too is a prognosticating science, and this type belongs neither to the theoretical nor to the practical sciences, although it is reputed to have a practical value. The telesmatical art is vain, for if we assume the positions of the spheres to exert a power on artificial products, this power will remain inside the product and not pass on to things outside it. As to conjuring, this is the type of thing that produces wonder, but it is certainly not a theoretical science. Whether alchemy really exists is very dubious; if it exists, its artificial product cannot be identical with the product of nature; art can at most become similar to nature but cannot attain nature itself in reality. As to the question whether it can produce anything which resembles the natural product generically, we do not possess sufficient data to assert categorically its impossibility or possibility, but only prolonged experiments over a lengthy period can procure the necessary evidence. I shall treat the four points Ghazali mentions one after the other.

Ghazali says :

The first point is their assertion that this connexion observed between causes and effects is of logical necessity, and that the existence of the cause without the effect or the effect without the cause is not within the realm of the contingent and possible. The second point is their assertion that human souls are substances existing by themselves, not imprinted on the body, and that the meaning of death is the end of their attachment to the body and the end of their direction of the body; and that otherwise the soul would exist at any time by itself. They affirm that this is known by demonstrative proof. The third point is their assertion that these souls cannot cease to exist, but that when they exist they are eternal and their annihilation cannot be conceived. The fourth point is their assertion that these souls cannot return to their bodies.

As to the first point, it is necessary to contest it, for on its negation depends the possibility of affirming the existence of miracles which interrupt the usual course of nature, like the changing of the rod into a serpent or the resurrection of the dead or the cleavage of the moon and those who consider the ordinary course of nature a logical necessity regard all this as impossible. They interpret the resurrection of the dead in the Koran by saying that the cessation of the death of ignorance is to be understood by it, and the rod which conceived the arch-deceiver, the serpent, by saying that it means the clear divine proof in the hands of Moses to refute the false doctrines of the heretics; and as to the cleavage of the moon they often deny that it took place and assert that it does not rest on a sound tradition; and the philosophers accept miracles that interrupt the usual course of nature only in three cases.

First: in respect to the imaginative faculty they say that when this faculty becomes predominant and strong, and the senses and perceptions do not submerge it, it observes the Indelible Tablet, and the forms of particular events which will happen in the future become imprinted on it; and that this happens to the prophets in a waking condition and to other people in sleep, and that this is a peculiar quality of the imaginative faculty in prophecy.

Secondly: in respect of a property of the rational speculative faculty i.e. intellectual acuteness, that is rapidity in passing from one known thing to another; for often when a problem which has been proved is mentioned to a keen-sighted man he is at once aware of its proof, and when the proof is mentioned to him he understands what is proved by himself, and in general when the middle term occurs to him he is at once aware of the conclusion, and when the two terms of the conclusion are present in his mind the middle term which connects the two terms of the conclusion occurs to him. And in this matter people are different; there are those who understand by themselves, those who understand when the slightest hint is given to them, and those who, being instructed, understand only after much trouble; and while on the one hand it may be assumed that incapacity to understand can reach such a degree that a man does not understand anything at all and has, although instructed, no disposition whatever to grasp the intelligibles, it may on the other hand be assumed that his capacity and proficiency may be so great as to arrive at a comprehension of all the intelligibles or the majority of them in the shortest and quickest time. And this difference exists quantitatively over all or certain problems, and qualitatively so that there is an excellence in quickness and easiness, and the understanding of a holy and pure soul may reach through its acuteness all intelligibles in the shortest time possible; and this is the soul of a prophet, who possesses a miraculous speculative faculty and so far as the intelligibles are concerned is not in need of a teacher; but it is as if he learned by himself, and he it is who is described by the words *the oil of which would well-nigh give light though no fire were in contact with it, light upon light*.

Thirdly: in respect to a practical psychological faculty which can reach such a pitch as to influence and subject the things of nature: for instance, when our soul imagines something the limbs and the potencies in these limbs obey it and move in the required direction which we imagine, so that when a man imagines something sweet of taste the corners of his mouth begin to water, and the potency which brings forth the saliva from the places where it is springs into action, and when coitus is imagined the copulative potency springs into action, and the penis extends; indeed, when a man walks on a plank between two walls over an empty space, his imagination is stirred by the possibility of falling and his body is impressed by this imagination and in fact he falls, but when this plank is on the earth, he walks over it without falling. This happens because the body and the bodily faculties are created to be subservient and subordinate to the soul, and there is a difference here according to the purity and the power of the souls. And it is not impossible that the power of the soul should reach such a degree that also the natural power of things outside a man's body obeys it, since the soul of man is not impressed on his body although there is created in man's nature a certain impulse and desire to govern his body. And if it is possible that the limbs of his body should obey him, it is not impossible that other things besides his body should obey him and that his soul should control the blasts of the wind or the downpour of rain, or the striking of a thunderbolt or the trembling of the earth, which causes a land to be swallowed up with its inhabitants. The same is the case with his influence in producing cold or warmth or a movement in the air; this warmth or cold comes about through his soul, all these things occur without any apparent physical cause, and such a thing will be a miracle brought about by a prophet. But this only happens in matters disposed to receive it, and cannot attain such a scale that wood could be changed into an animal or that the moon, which cannot undergo cleavage, could be cloven. This is their theory of miracles, and we do not deny anything they have mentioned, and that such things happen to prophets; we are only opposed to their limiting themselves to this, and to their denial of the possibility that a stick might change into a serpent, and of the resurrection of the dead and other things. We must occupy ourselves with this question in order to be able to assert the existence of miracles and for still another reason, namely to give effective support to the doctrine on which the Muslims base their belief that God can do anything. And let us now fulfill our intention.

The ancient philosophers did not discuss the problem of miracles, since according to them such things must not be examined and questioned; for they are the principles of the religions, and the man who inquires into them and doubts them merits punishment, like the man who examines the other general religious principles, such as whether God exists or blessedness or the virtues. For the existence of all these cannot be doubted, and the mode of their existence is something divine which human apprehension cannot attain. The reason for this is that these are the principles of the acts through which man becomes virtuous, and that one can only attain knowledge after the attainment of virtue. One must not investigate the principles which cause virtue before the attainment of virtue, and since the theoretical sciences can only be perfected through assumptions and axioms which the learner accepts in the first place, this must be still more the case with the practical sciences.

As to what Ghazali relates of the causes of this as they are according to the philosophers, I do not know anyone who asserts this but Avicenna. And if such facts are verified and it is possible that a body could be changed qualitatively through something which is neither a body nor a bodily potency, then the reasons he mentions for this are possible; but not everything which in its nature is possible can be done by man, for what is possible to man is well known. Most things which are possible in themselves are impossible for man, and what is true of the prophet, that he can interrupt the ordinary course of nature, is impossible for man, but possible in itself; and because of this one need not assume that things logically impossible are possible for the prophets, and if you observe those miracles whose existence is confirmed, you will find that they are of this kind. The clearest of miracles is the Venerable Book of Allah, the existence of which is not an interruption of the course of nature assumed by tradition, like the changing of a rod into a serpent, but its miraculous nature is established by way of perception and consideration for every man who has been or who will be till the day of resurrection. And so this miracle is far superior to all others.

Let this suffice for the man who is not satisfied with passing this problem over in silence, and may he understand that the argument on which the learned base their belief in the prophets is another, to which Ghazali himself has drawn attention in another place, namely the act which proceeds from that quality through which the prophet is called prophet, that is the act of making known the mysterious and establishing religious laws which are in accordance with the truth and which bring about acts that will determine the happiness of the totality of mankind. I do not know anyone but Avicenna who has held the theory about dreams Ghazali mentions. The ancient philosophers assert about revelation and dreams only that they proceed from God through the intermediation of a spiritual incorporeal being which is according to them the bestower of the human intellect, and which is called by the best authors the active intellect and in the Holy Law angel.

Annotated Bibliography

Ahmad, S. Maqbul. 1981. "Al-Masudi." In *Dictionary of Scientific Biography* (hereafter cited as *DSB*), 9: 171–172.

———. 1991. "Djughrafiya." In *Encyclopaedia of Islam* (hereafter cited as *EI*), 3: 575–587.

———. 1997. "Kharita." *EI*, 4: 1077–1083.

Aristotle. 1984. *The Complete Works of Aristotle. The Revised Oxford Translation*. Jonathan Barnes, ed. 2 vols. Princeton, NJ: Princeton University Press.

Arnaldez, Roger. 2000. *Averroes: A Rationalist in Islam*. (Translated from the French original, *Averroès, un rationaliste en Islam*, first published by Éditions Balland, Paris, 1998, by David Streight.) Notre Dame: University of Notre Dame Press. An insightful short book of 156 pages on the thought of Ibn Rushd by a member of the Académie Française and professor emeritus at the Sorbonne.

al-Attas, Syed Muhammad Naquib. 1981. *The Positive Aspects of Tasawwuf: Preliminary Thoughts on an Islamic Philosophy of Science*. Kuala Lumpur: Islamic Academy of Science. This short book by a contemporary Muslim scholar outlines the foundational principles of science from an Islamic perspective.

———. 1989. *Islam and the Philosophy of Science*. Kuala Lumpur: International Institute of Islamic Thought and Civilization.

Azami, M. M. 1978. *Studies in Early Hadith Literature*. Indianapolis: American Trust Publications. A pioneering work on the evaluation of the smaller collections of *Hadith* antedating the six canonical collections, based on a Ph.D. thesis that has been called by Professor A. J. Arberry "the most exciting and original investigations in this field."

———. 2003. *The History of the Qur'anic Text: From Revelation to Compilation*. Leicester: UK Islamic Academy. A remarkable work, one of the best works in English, on the history of the Qur'anic text. It begins with a brief history of Islam and then details the coming into existence of the Qur'an—from its revelation to its compilation in the form of a book. Based on the early original sources, this work provides numerous insights into the procedures through which the text of the Qur'an was preserved.

Bacon, Francis. 1905. *The Philosophical Works of Francis Bacon*. John M. Robertson, ed. London: Routledge.

Baljon, J. M. S. 1961. *Modern Muslim Koran Interpretation, 1880–1960*. Leiden: Brill.

Banu Musa. ca. 9th century. *Kitab al-Hiyal*. Translated in 1979 by Donald Hill as *The Book of Ingenious Devices*. Dordrecht: D. Reidel. This translation provides an opportunity to understand some of the lesser-known aspects of Islamic scientific tradition dealing with technological devices used by Muslims before the emergence of modern technology.

Barbour, Ian. 2000. *When Science Meets Religion*. San Francisco: HarperSanFrancisco. An important work in the contemporary religion and science discourse. Barbour's view of science and religion is, however, totally based on his particular Christian understanding of the issues. As such, his schema is applicable to a small segment of the science and religion discourse.

———. 2002. "Response: Ian Barbour on Typologies." *Zygon*, 37: 345–359.

Berggren. J. L. 1986. *Episodes in the Mathematics of Medieval Islam*. New York: Springer-Verlag. A major contribution to our understanding of the history of mathematics, with 97 figures and 20 plates.

———. 1996. "Islamic Acquisition of the Foreign Sciences: A Cultural Perspective." In Jamil Ragep and Sally Ragep, eds., *Tradition, Transmission, Transformation*. Leiden: Brill, pp. 263–283. This is a revised version of an earlier paper, which appeared in *The American Journal of Islamic Social Sciences (AJISS)*, 1992; 9(3): 310–324.

al-Biruni, Abu Rayhan. Tr. 1879. *Al Athar al Baqiya*. Translated by Edward Sachau as *The Chronology of Ancient Nations*. London: W. H. Allen and Co.

———.Tr. 1967. *Kitab Tahdid Nihayat al-Amakin Litashih Masafat al-Masakin*. Translated by Jamil Ali as *The Determination of the Coordinates of Positions for the Correction of Distances between Cities*. Beirut: The American University of Beirut. This translation of one of the most important eleventh-century works on geography also provides insights into the nature of Islamic scientific tradition.

———. Tr. 1989. *Kitab al-Jamahir fi Marifatil Jawahir*. Translated by Hakim Mohammad Said as *The Book Most Comprehensive in Knowledge on Precious Stones*. Islamabad: Pakistan Hijrah Council.

———. Tr. 2001–2006. *Al-As'ilah wa'l Ajwibah*. Translated by Rafik Berjak and Muzaffar Iqbal as *Ibn Sina–al-Biruni Correspondence*. In *Islam and Science*, Vols. 1–4. This is the translation of the correspondence between Ibn Sina and al-Biruni from the original Arabic text, edited with English and Persian introductions by Seyyed Hossein Nasr and Mehdi Mohaghegh. Kuala Lumpur: International Institute of Islamic Thought and Civilization, 1995.

———. Tr. 1976. *Afrad'l Maqal fi amr al-zilal*. Translated with commentary by E. S. Kennedy as *The Exhaustive Treatise on Shadows*. 2 vols. Aleppo: Institute for the History of Arabic Science. This is an exhaustive discussion of shadows, their nature, properties, and utilities. The work deals with optics, mathematics, astronomy, religion, etymology, and literature. It has direct relevance to the determination of prayer times.

Bucaille, Maurice. 1976. *La Bible, le Coran et la science: les Écritures saintes examinées à la lumière des connaissances modernes*. Paris: Seghers. Translated into English by Alastair D. Pannell and the author as *The Bible, the Qur'an and Science: The Holy Scriptures Examined in the Light of Modern Knowledge*. Indianapolis: North American Trust Publications, 1978. This is the most popular book on Islam and science in the Muslim world. It has been translated into every major Islamic language and millions of copies have sold. It inaugurated a whole branch of facile writings on Islam and science which attempts to read modern science into the Qur'an and *Sunnah*.

Campbell, William. 1986. *The Qur'an and the Bible In The Light of History and Science*. Upper Darby, PA: Middle East Resources. A response to Maurice Bucaille's book, *The Bible, the Qur'an and Science: The Holy Scriptures Examined in the Light of Modern Knowledge*.

Cantor, Geoffery and Kenny, Chris. 2001. "Barbour's Fourfold Way: Problems with His Taxonomy of Science-Religion Relationships." *Zygon*, 36: 765–781.

Ceylan, Yasin. 1996. *Theology and Tafsir in the Major Works of Fakhr al-Din al-Razi*. Kuala Lumpur: International Institute of Islamic Thought.

Chabás, José and Goldstein, Bernard R. 2003. *The Alfonsine Tables of Toledo*. Dordrecht: Kluwer Academic Press.

Chittick, William C. 1989. *The Sufi Path of Knowledge*. Albany: State University of New York Press. A groundbreaking work on Ibn al-Arabi. Chittick's book is a significant contemporary contribution to Islamic mysticism.

———. 1998. *The Self-Disclosure of God: Principles of Ibn al-'Arabi's Cosmology*. Albany: State University of New York Press. This book deals with Ibn al-Arabi's cosmology. It provides insights into later Islamic mysticism and reformulation of philosophical cosmologies.

Cohen, Floris H. 1994. *The Scientific Revolution: A Historical Inquiry*. Chicago: University of Chicago Press.

Craig, William Lane. 1979. *The Kalam Cosmological Argument*. New York: The Macmillan Press Ltd. One of the best descriptions of the *Kalam* cosmological argument, in two parts. The first part deals with the thought of al-Kindi, Saadia, and al-Ghazali; the second with modern defence of the *Kalam* cosmological argument. The book includes two appendices, which describe (i) the relationship between the *Kalam* cosmological argument and (ii) Zeno's paradox, and *Kalam* cosmological argument and the thesis of Kant's first antinomy.

Daniel, Norman. 1960, reprint. 2000. *Islam and the West: The Making of an Image*. Oxford: Oneworld. A pioneering work on the making of image of Islam in the West.

Dante, Alighieri. 1971. *The Divine Comedy, Inferno*. Translated by Mark Musa. New York: Penguin.

Darwin, Charles. 1887. *The Descent of Man and Selection in Relation to Sex*. Reprinted 1998, New York: Prometheus Books.

Davidson, Herbert A. 1987. *Proofs for Eternity, Creation and the Existence of God in Medieval Islamic and Jewish Philosophy*. Los Angeles: Oxford University Press.

One of the most exhaustive studies on the issue of eternity and creation, providing insights into the making of the debate during the Middle Ages. Davidson provides a comprehensive survey of the views of major Muslim and Jewish philosophers along with the history of the debate.

Dembski, William, ed. 1998. *Mere Creation.* Downers Grove: InterVarsity Press.

al-Dhahabi, Muhammad Husayn. 1965 [reprint 1985]. *al-Tafsir wa'l-Mufassirun,* 2 vols, 4th ed. Cairo: Maktabat al-Wahba. A useful survey of Tafsir literature. Al-Dhahabi was one of the first scholars to include scientific *tafsir* in his survey.

Dhanani, Alnoor. 1994. *The Physical Theory of Kalam.* Leiden: Brill.

———.1996. "Kalam Atoms and Epicurean Minimal Parts." In Jamil F. Ragep and Sally P. Ragep, eds. *Tradition, Transmission, Transformation.* Leiden: Brill, pp. 157–171.

Dictionary of Scientific Biography (cited as *DSB* 1980). 1970–1980. Charles C. Gillispie, ed. 16 vols. New York: Charles Scribner's Sons. A major reference work and one of the few contemporary sources wherein Muslim scientists are placed in the general history of science.

Dizer, Muammar. 2001. "Observatories and Astronomical Instruments." In A. Y. Al-Hassan, ed., *The Different Aspects of Islamic Culture.* Vol 4: Science and Technology in Islam. Part 1: The Exact Sciences. Paris: UNESCO Publishing, pp. 235–265. A compilation of articles by major scholars in the history of science, this is a good secondary work on specific aspects of Islamic science.

Encyclopedia of Islam (New Edition). 1997. 11 vols. Leiden: Brill. This crown in the jewel of orientalism remains the most used reference work in Islamic studies. Most of the articles are written by orientalists and lack authenticity from the point of view of Muslims.

The Encyclopedia of Religion and Nature. 2005. Bron R. Taylor, ed. 2 vols. New York: Thoemmes Continuum.

al-Faruqi, Ismail, R. 1982. *Islamization of Knowledge: General Principles and Work Plan.* Washington DC: International Institute of Islamic Thought. This book has been republished in a revised and enlarged edition by a group of scholars associated with the International Institute of Islamic Thought. The Institute also publishes a quarterly, *Journal of Islamic Social Sciences.*

Feingold, Mordechai. 1996. "Decline and Fall: Arabic Science in Seventeenth Century England." In Ragep and Ragep eds., *Tradition, Transmission, Transformation.* Leiden: Brill, pp. 441–469.

al-Ghazali, Abu Hamid. Tr. 2000. *Tahafut al-falasifa.* Translated by Michael E. Marmura as *The Incoherence of the Philosophers.* Provo: Brigham Young University Press. One of the most important works of al-Ghazali, which is frequently erroneously regarded as the book that destroyed science in Islamic civilization. This translation by a well-known Ghazali scholar provides a useful parallel English-Arabic text and annotation.

Goldberg, Hillel. 2000. "Genesis, Cosmology and Evolution." In *Jewish Action,* 5769: 2.

Goldziher, Ignaz. 1915. "Stellung der alten islamischen Orthodoxie zu den antiken Wissenschaften," *Abhandl. Der Preuss. Akad. D. Wiss. (Philos.-hist. Kl.)*

8: 3–46; English translation as "The Attitude of Orthodox Islam Toward the 'Ancient Sciences'" in Merlin L. Swartz, ed., 1981. *Studies on Islam*. New York: Oxford University Press, pp. 185–215. A pioneering but biased study on the relationship between Islam and natural sciences that has influenced the course of scholarship on Islam and science for almost a century.

Grant, Edward. 2004. *Science and Religion, 400 B.C. to A.D. 1550: From Aristotle to Copernicus*. Westport: Greenwood Press. A general survey of interaction of science and Christianity, with a small section devoted to Islam.

Greaves, Richard L. 1969. *The Puritan Revolution and Educational Thought*. Piscataway, NJ: Rutgers University Press.

Guénon, René. *The Crisis of the Modern World*. 1942. London: Luzac & Co. This is the English translation by Arthur Osborne of Guénon's French work, *La crise du monde moderne*. Guénon pioneered traditional studies in modern times. His work has influenced discourse on Islam and science in foundational ways.

Gutas, Dimitri. 1998. *Greek Thought, Arabic Culture: Graeco-Arabic Translation Movement in Baghdad and Early Abbasid Society*. London: Routledge. This detailed and well-referenced study of the Graeco-Arabic translation movement brings into full relief a large amount of research on the subject. Its extensive references to original sources provide valuable information about the patrons and translators who sustained this movement for almost three centuries. It analyzes various theories that have been proposed to explain this movement with scholarly vigor and rejects some of the most commonly cited reasons for the emergence of this movement. With all its academic strength, the work remains deeply entrenched in orientalism and views Islam and science nexus from that perspective.

———. 2003. "Islam and Science: A False Statement of the Problem." *Islam & Science*, 1(2): 215–220.

Hall, Roberts. 2001. "Mechanics." In A. Y. Al-Hassan, ed., *The Different Aspects of Islamic Culture*. Vol. 4: Science and Technology in Islam. Part 1: The Exact Sciences. Paris: UNESCO Publishing, pp. 297–336.

al-Hamawi, Yaqut. Tr. 1959. *Al-Mua'jam al-Buldan*. The introductory chapter, translated by Wadie Jwaideh as *The Introductory Chapters of Yaqut's Mua'jam al-Buldan*. Leiden: Brill.

Haq, Nomanul Syed. 1994. *Names, Nature and Things*. Dordrecht: Kluwer Academic Publishers. Based on a doctoral thesis, this book is a major contribution to our understanding of the work of Jabbir bin Hayyan, one of the most important scientists in the history of science.

———. 2001. "Islam." In Dale Jamieson, ed., *A Companion to Environmental Philosophy*. Oxford: Blackwell Publishers, pp. 111–129.

Heisenberg, Werner. 1958. *Physics and Philosophy: The Revolution in Modern Science*. New York: Prometheus Books.

Hodgson, Marshall. 1974. *The Venture of Islam*. Chicago: University of Chicago Press. A major study on Islamic civilization.

Hoodbhoy, Pervez. 1991. *Islam and Science: Religious Orthodoxy and the Battle for Rationality*. London and New Jersey: Zed Books Ltd.

Huff, Toby E. 1993. *The Rise of Early Modern Science: Islam, China and the West.* Cambridge: University of Cambridge. An often-quoted work that attempts to show that there is something inherently wrong with Islam, and that therefore it could not have produced science. A typical orientalist approach that attempts to deconstruct Islam and its tradition of learning.

Ibn Kathir. reprint 1998. *Tafsir ul-Qur'an al-'Azim,* Sami b. Muhammad as-Salamah, ed. 8 vols. Riyad: Dar Tayyibah. This is one of the major commentaries of the Qur'an, written in the fourteenth century, and still widely used.

Ibn al-Nadim. Tr. 1970. *Al-Fihrist.* Translated by Bayard Dodge as *The Fihrist.* Cambridge: Columbia University Press. One of the best sources of information on a wide range of subjects, Ibn Nadim's *Fihrist* remains a major reference work on various aspects of Islamic tradition of learning—from authors to biographies to subject areas.

Ibn Rushd, Abu al-Waleed Muhammad. Tr. 1954. *Tahafut al-Tahafut (The Incoherence of the Incoherence).* Translated from the Arabic with Introduction and Notes by Simon van den Bergh, 2 vols. London: Luzac. A key work for Islam and science discourse, Ibn Rushd here refutes al-Ghazali's major work, *The Incoherence of Philosophers.*

Ibn Rushd, Abu al-Waleed. 1959. *Kitab fasl al-maqal,* Arabic text, edited by George F. Hourani. Leiden: Brill. Often cited as a proof of Islam's supposed antagonistic attitude toward natural sciences in secondary works, this important book by Ibn Rushd is actually concerned with investigating "parentage" (*ittisal*) between Islamic religious law (*shariah*) and wisdom.

Ibn Sina. Tr. 1930 [reprint 1999]. *The Canon of Medicine. A treatise on the Canon of medicine of Avicenna, incorporating a translation of the first book, by O. Cameron Gruner.* London: Luzac. This major eleventh-century work remained the standard textbook for teaching medicine both in the Muslim world and in Europe for several centuries.

Ibn Sina. Tr. 1974. *The Life of Ibn Sina.* A critical edition and annotated translation by William E. Gohlman. Albany: State University of New York Press.

Ibn Sina, al-Biruni. 2003. *al-As'ilah wa'l-Ajwibah.* Translated by Rafik Berjak and Muzaffar Iqbal, *Islam & Science,* 1(1–2): 91–98, 253–260.

Ibn Tufayl, Abu Bakr Muhammad. 1936 [reprint 1972]. *The Journey of the Soul,* A new translation by Riad Kocache. London: Octagon. A short book which attempted to show that, when left to its own, human reason can arrive at fundamental truths about God and life.

Inalcik, Halil and Quataert, Donald, eds., 1994. *An Economic and Social History of the Ottoman Empire.* 2 vols. Cambridge: Cambridge University Press.

Inati, Shams. 1996. "Ibn Sina." In *History of Islamic Philosophy.* Seyyed Hossein Nasr and Oliver Leaman, eds. London: Routledge, pp. 231–246.

Iqbal, Muzaffar. 2002. *Islam and Science.* Aldershot: Ashgate.

al-Iskandarani, Muhammad b. Ahmad. 1880. *Kashf al-Asrar 'an al-Nuraniyya al-Qur'aniyya fima yataallaqu bi'l-Ajram as-Samawiyya wa'l-Ardiyya wa'l-Haywanat wa'l-Nabat wa'l-Jawahir al-Madaniyya,* 3 vols. Cairo: Maktabat al-Wahba. The first complete scientific commentary on the Qur'an, this work inaugurated a new branch in the centuries-old tradition of *tafsir* (commentary). It attempts to show many discoveries of modern science in the Qur'an.

———. 1300/1883. *Tibyan al-Asrar al-Rabbaniyya fi'l-Nabat wa'l Ma'din wa'l-Khawass al-Haywaniyya* (*The Demonstration of Divine Secrets in the Vegetation and Minerals and in the Characteristics of Animals*). Damascus: n. p.,. The word *tibyan* (explanation) in the title is taken from Q. 16:89: *And We have sent down to thee the Book, explaining all things—a guide, a mercy and glad tidings for Muslims*. A work that attempts to show that many facts and theories of modern science are already in the Qur'an.

Izutsu, Toshihiko. 1964. *God and Man in the Koran: Semantics of the Qur'anic Weltanschauung*. Tokyo: The Keio Institute of Cultural and Linguistic Studies, reprinted 2002 as *God and Man in the Qur'an: Semantics of the Qur'anic Weltanschauung*. Kuala Lumpur: Islamic Book Trust. An important work on the semantics of the Qur'an by a Japanese scholar of Islam that examines the semantic structure of the Qur'an and the mutual relationship of its essential themes. All references in the text are to the 2002 edition.

Jansen, J. J. G. 1974. *The Interpretation of the Koran in Modern Egypt*. Leiden: Brill. A general survey of Qur'an scholarship in Egypt.

al-Jawziyya, Ibn Qayyim. Tr. 1998. *Medicine of the Prophet*. Translated by Penelope Johnstone. Cambridge: Islamic Texts Society. A well-produced translation of the classical work of Ibn Qayyim al-Jawziyya (1292–1350), *al-Tibb al-Nabawi*. The book contains a useful "Introduction" by the translator, outlining the theoretical foundation of the Prophetic medicine and brief notes on prominent Muslim physicians.

al-Jazzari, Abu'l Izz ibn al-Razzaz. Tr. 1974. *The Book of Knowledge of Ingenious Mechanical Devices*. Translated by Donald R. Hill. Dordrecht: D. Reidel.

Kalin, Ibrahim. 2001. "The Sacred Versus the Secular: Nasr on Science." In Lewis Edwin Hahn, Randall E. Auxier, and Lucian W. Stone, Jr., eds., *The Philosophy of Seyyed Hossein Nasr*. Peru, Ill: Open Court, pp. 445–462.

———. 2002. "Three Views of Science in the Islamic World." In Ted Peters, Muzaffar Iqbal, and Syed Nomanul Haq, eds., *God, Life, and the Cosmos*. Aldershot: Ashgate, pp. 47–76.

Kamali, Mohammad Hashim. 2005. *A Textbook of Hadith Studies*. Leicestershire: Islamic Foundation. An introductory book on Hadith. Includes Hadith criticism, classification, compilation, and authenticity. This book provides a wide-ranging coverage of Hadith methodology and literature for introductory and intermediate levels.

Keddie. Nikki, R. 1968. *An Islamic Response to Imperialism*. Berkeley: University of California Press. A good source on the life and times of Jamal al-Din al-Afghani.

———. 1972. *Sayyid Jamal ad-Din al-Afghani: A Political Biography*. Berkeley: University of California.

Kennedy, E. S. 1960. *The Planetary Equatorium of Jamshid Ghiyath al-Din al-Kashi*. An annotated translation and commentary on the construction and use of two astronomical computing instruments invented by the fifteenth century Iranian scientist,

Jamshid ibn Masud ibn Mahmud, Ghiyath al-Din al-Kashi (d. 1429). Princeton: Princeton University Press. This short book provides insights into the natural of Islamic science in the fifteenth century; it helped in the reexamination of the question of decline of science in Islamic civilization and actually pushed the date of the so-called decline to a period after the fifteenth century.

———. (1967). "The Lunar Visibility Theory of Yaqub ibn Ôariq." *Journal of Near Eastern Studies*, 27: 126–132.

Kennedy, E. S. et al. 1983. *Studies in the Islamic Exact Sciences*. Beirut: American University of Beirut.

———. 1980. "Al-Biruni." *DSB*, 1: 147–158.

Khan, Sayyid Ahmad. 1961. *Musafran-e London*. Lahore: Majlis Taraqi-e Adab.

———. 1963. *Maqalat-e* Sir Sayyid. Maulana Muhammad Ismail Panipati, eds. 16 vols. Lahore: Majlis-e Taraqqi-e Adad.

al-Khwarizmi, Abu Ja'far Muhammad b. Musa. Reprint. 1989. *Kitab al-Mukhtasar fi'l-hisab al-jabr wa'l muqablah*. Islamabad: Hijrah Council. This book inaugurated the science of Algebra. Al-Khwarizmi's treatise is now of historical significance, but as such it remains an important document.

al-Kindi, Ya'qub ibn Ishaq. *On First Philosophy (fi al-Falsafah al-Ula)*. Translated by Alfred L. Ivry, Albany: State University of New York Press. This book provides a good insight into al-Kindi's philosophy through a representative work.

King, David. 1999. *World-Maps for Finding the Direction and Distance to Mecca*. Leiden: Brill. The book describes two newly discovered instruments of the seventeenth century that were used to determine direction and distance to Makkah. In addition, it provides a comprehensive survey of literature on the subject.

———. 2004. *In Synchrony with the Heavens: Studies in Astronomical Timekeeping and Instrumentation in Medieval Islamic Civilization*. 2 vols. Leiden: Brill. A monumental work of approximately 2000 pages, based on a large number of sources that have not previously been studied.

Kraus, Paul. 1991. "Djabir b. Hayyan" *EI*, 2: 257–359.

Lamoureux, Denis O., Phillip E. Johnson, et al. 1999. *Darwinism Defeated?* Vancouver: Regent College Publishing.

Lattin, Harriet Pratt. Tr. 1961. *The Letters of Gilbert with His Papal Privileges as Sylvester II*. New York: Columbia University Press.

Lemay, Richard. 1981. "Gerard of Cremona." *DSB*, 5: 173–192.

Lewis, Bernard. 1961. *The Emergence of Modern Turkey*. London: Oxford University Press.

Lindberg, David C. 1992. *The Beginnings of Western Science*. Chicago: The University of Chicago Press.

Lings, Martin. 1980. *Ancient Beliefs and Modern Superstitions*. London: Perennial Books. An examination of modernity from the traditional perspective.

———. 1988. *The Eleventh Hour*. Cambridge: Archetype. The book has an important section on the theory of evolution.

———. 2004. *Mecca from Before Genesis until Now*. Cambridge: Archetype.

Lloyd, G. E. R. 1973. *Greek Science after Aristotle*. New York: W. W. Norton. A short book which provides a good overall picture of Greek science after Aristotle.

McVaugh, Michael. 1981. "Constantine the African." *DSB*, 3: 393–395.

Makdisi, George. 1991. *Religion, Law and Learning in Classical Islam*. Aldershot: Variorum. A pioneering work which has facilitated a better understanding of the institutions of learning in Islam.

Malik, Hafeez. 1980. *Sir Sayyid Ahmad Khan and Muslim Modernism in India and Pakistan*. New York: Columbia University Press.

Mardin, Serif. 2000. *The Genesis of Young Ottoman Thought: A Study in the Modernization of Turkish Political Ideas*. Syracuse: Syracuse University Press.

Mohammad, A. H. Helmy. 2000. "Notes on the Reception of Darwinism in Some Islamic Countries." In E. Ihsanoğlu and F. Günergun, eds., *Science in Islamic Civilization*. Istanbul: IRCICA, pp. 245–255.

Moore, Keith L. 1982. *The Developing Human: With Islamic Additions*. Jeddah: Commission for Scientific Miracles of Qur'an and Sunnah. A popular work which attempts to show a relationship between the Qur'anic description of the development of fetus and current scientific understanding.

———. 1993. *Qur'an and Modern Science: Correlation Studies*. Jeddah: Islamic Academy for Scientific Research, Jeddah.

Moosa, Ebrahim. 2002. "Interface of Science and Jurisprudence: Dissonant Gazes at the Body in Modern Muslim Ethics." In Ted Peters, Muzaffar Iqbal, and Syed Nomanul Haq, eds., *God, Life, and the Cosmos*. Aldershot: Ashgate.

Nasr, Seyyed Hossein. 1968. *Science and Civilization in Islam*. Cambridge: Harvard University Press. One of the first well-researched books in modern times on science and Islamic civilization. The book was a pioneering effort in showing various internal connections between Islam and its scientific tradition. It also presented a large amount of new information on Muslim scientists and their works to the general readership.

———. 1981. *Islamic Life and Thought*. Albany: State University of New York Press.

———. 1993. *An Introduction to Islamic Cosmological Doctrines*, rev. ed. Albany: State University of New York Press. One of the first comprehensive accounts of Islamic philosophical cosmologies written in modern times. The book contains chapters on cosmological doctrines of Ibn Sina, Ikhwan al-Safa, and al-Biruni.

———. 1993. *The Need for a Sacred Science*. Albany: State University of New York Press.

———. 1996. *Religion and the Order of Nature*. New York: Oxford University Press. A highly well documented account of views of the various religious traditions about nature. Based on his 1994 Cadbury Lectures at the University of Birmingham, the book presents a wide spectrum of beliefs regarding the order of nature.

———. 1997. *Sadr al-Din Shirazi and his Transcendent Theosophy*. Tehran: Institute for Humanities and Cultural Studies. A short, but useful, work on Mulla Sadra containing background to his contribution to Islamic philosophy, a short biography, and description of his works.

———. 2001a. "Cosmology." In A. Y. al-Hassan, ed., *Different Aspects of Islamic Culture*. 2 parts. Paris: UNESCO Publishing.

———. 2001b. "An Intellectual Biography." In Lewis Edwin Hahn, Randall E. Auxier, and Lucian W. Stone, Jr., eds., *The Philosophy of Seyyed Hossein Nasr*. Peru, Ill: Open Court.

———. 2001c. "Reply to Ibrahim Kalin." In Lewis Edwin Hahn, Randall E. Auxier, and Lucian W. Stone, Jr., eds., *The Philosophy of Seyyed Hossein Nasr*. Peru, Ill: Open Court, pp. 463–468.

Northbourne, Lord. 1983. *Looking Back on Progress*. Lahore: Suhail Academy.

Nursi, Badiuzzaman. 1998. *The Words: On the Nature of Man, Life and All Things*. Istanbul: Sozler Publications. A collection of articles and speeches by Said Nursi.

Olson, Richard G. 2004. *Science and Religion, 1450–1900. From Copernicus to Darwin*. Westport: Greenwood Press. Part of the Greenwood Guides to Science and Religion, this book outlines the nature of the interaction between science and Christianity between 1450 and 1900.

Owen, Roger. 2004. *Lord Cromer: Victorian Imperialist, Edwardian Proconsul*. Oxford and New York: Oxford University Press. The first biography of the British de facto ruler of Egypt from 1883 to 1907.

Pagel, Walter. 1977. "Medical Humanism: A Historical Necessity in the Era of the Renaissance." In Francis Maddison, Margaret Pelling, and Charles Webster, eds. *Linacre Studies: Essays on the Life and Works of Thomas Linacre c. 1460–1525*. Oxford: Clarendon Press, pp. 375–386.

Panipati, Shaikh Muhammad Ismail. ed., 1993. *Letters to and from Sir Syed Ahmad Khan*. Lahore: Majlis Taraqi-e Adab, Board for Advancement of Literature. These letters are an important source for understanding the worldview of Indian reformer Sayyid Ahmad Khan, who wrote a partial commentary of the Qur'an in which he attempted to show a relationship between modern science and the Qur'an.

Pearson, J. D. 1986. "Al-Kur'an." *EI*, 5: 400–432.

Perry, Whitall N., 1995. *The Widening Breach: Evolutionism in the Mirror of Cosmology*. Barlow: Quinta Essentia. A traditional critique of the theory of evolution, this is by far the best work which examines the theory of evolution from the perspective of cosmology.

Pines, Shlomo. 1986. "What was original in Arabic science." In *Studies in Arabic Versions of Greek Texts and in Medieval Science*. Jerusalem: Magnes, the Hebrew University, p. 193.

Pingree, David. 1970. "The Fragments of the Works of al-Fazari." *Journal of Near Eastern Studies*. 29: 103–123.

Pratt, Lattin. 1961. *The Letters of Gilbert with His Papal Privileges as Sylvester II*. Translated with an Introduction. New York: Columbia University Press.

Q. Al-Qur'an. Muslims consider the Qur'an to be the actual speech of God and hence untranslatable; only the meanings of the verses can be communicated in another language. Most "translations," therefore, are renderings of the Arabic text. In quoting the verses of the Qur'an, the first number refers to the *surah* (chapter), the second to the *ayah* (verse).

Qurashi, M. M., Bhutta, S. M., and Jafar, S. M. 1987. *Quranic Ayaat Containing References to Science and Technology*. Islamabad: Sh. Sirri Welfare & Cultural Trust and Pakistan Science Foundation.

Rashed, Roshdi. 1994. *The Development of Arabic Mathematics: Between Arithmetic and Algebra*. Dordrecht: Kluwer Academic Publishers. This is the English translation by A. F. W. Armstrong of Rashed's 1984 work on mathematical sciences in Islam, *Entre Arithmétique et Algèbre. Recherches sur l'Histoire des Mathèmatiques Arabes*. Paris: Société d'Eition Les Belles Lettres.

Ratzsch, Del. 2000. *Science & Its Limits*. Downers Grove: InterVarsity Press. A book written from a Christian perspective that aspires to define science and its limits: what it can and cannot tell us.

Roberts, Victor. 1966. "The Planetary Theory of Ibn al-Shatir." *Isis*, 57: 210.

Sabra, A. and Shehaby, N. 1971. *Ibn al-Haytham al-Shukuk Ala Batlamyus (Dubitationes in Ptolemaeum)*. Cairo: National Library Press.

Sabra, A. I. 1987. "The Appropriation and Subsequent Naturalization of Greek Science in Medieval Islam: A Preliminary Statement." *History of Science*, 25: 223–243; reprinted with other papers in a collection, *Optics, Astronomy and Logic: Studies in Arabic Science and Philosophy*. 1994.

———. 1994. *Optics, Astronomy and Logic: Studies in Arabic Science and Philosophy*. Aldershot: Variorum.

Sabzavari, Haji Mulla Hadi. Tr. 1977. *The Metaphysics of Sabzavari*. Translated by Mehdi Mohaghegh and Izutsu Toshihiko. New York: Caravan Books. This work is a complete translation of the *Metaphysics* of Sabzavari's *Ghurar al-farâ'd*, a systematic exposition of traditional Islamic philosophy comprising logic, physics, theology, and metaphysics, commonly known as *Sharh-I manzumah (Commentary on a Philosophical Poem)*. This is the most popular textbook that has been used in Iran, and that is still in current use, to teach philosophy in almost all religious schools.

Saliba, George. 1994. *A History of Arabic Astronomy*. New York: New York University Press. A collection of articles, written over a period of two decades, devoted to the study of the various aspects of non-Ptolemaic astronomy in medieval Islam. Based on primary sources, these articles provide insights into the making of the Islamic astronomical tradition between the eleventh and the fifteenth centuries. Among other things, the book provides substantial evidence against the commonly held belief that Islamic scientific tradition declined as early as the twelfth century. The author shows that some of the techniques and mathematical theorems developed during this period were identical to those employed by Copernicus in developing his own non-Ptolemaic astronomy.

Samsó, Julio. 2001. "Astronomical Tables and Theory." In A. Y. Al-Hassan, ed., *The Different Aspects of Islamic Culture*. Vol. 4: Science and Technology in Islam. Part 1: The Exact Sciences. Paris: UNESCO, pp. 209–234.

Sardar, Ziauddin. ed., 1984. *The Touch of Midas: Science, Values and the Environment in Islam and the West*. Manchester: University of Manchester.

———. 1985. *Islamic Futures*. London: Mansell.

———. 1989. *Explorations in Islamic Science*. London: Mansell. Sardar's ideas on Islam and science were presented in this work in more detail than in his early work. Criticizing modern science from a sociological aspect, Sardar developed a new branch of Islam and science discourse using the work of Western critics of modern science.

Sarton, George. 1927–1947. *Introduction to the History of Science*. 3 vols. Baltimore: Williams and Wilkins. A major reference work of its times, this book brought to attention major achievements of Islamic scientific tradition. It has now become dated, but remains important for its historical role.

———. 1931–48. *Introduction to the History of Science*, 3 vols. Baltimore: Published for the Carnegie Institute of Washington by The Williams and Wilkins Company.

———. 1955. *Appreciation of Ancient and Medieval Science during the Renaissance (1450–1600)*. New York: Barnes.

Saunders, J. J. 1963. "The Problem of Islamic Decadence." *Journal of World History*, 7: 701–720.

as-Sawi, Abdul Jawwad. 1992. *Proposed Medical Research Projects Derived from the Qur'an and Sunnah*. Jeddah: Commission for Scientific Miracles of Qur'an and Sunna.

Sayili, Aydin. 1960. *The Observatory in Islam*. Ankara: Turk Tarih Kurumu Basimevi. One of the most important works on the subject, providing a systematic account of the observatories in ten chapters. An appendix contains reflections on the causes of the decline of science in the Muslim world.

Schuon, Frithjof. 1965. *Lights on the Ancient Worlds*. London: Perennial Books. This is the English translation by Lord Northbourne of Schuon's *Regards sur les mondes anciens*.

Setia, Adi. 2003. "Al-Attas' Philosophy of Science: An Extended Outline." *Islam & Science*, 1(2): 165–214.

Sharif, M. M. 1963. *A History of Muslim Philosophy*. 2 vols. Karachi: Royal Book Company. A major work on Islamic philosophy, contains biographical and bibliographic details and life sketches of major Muslim scholars.

al-SharqĀwą, Muṇammad ĂIffat. 1972. *IttijĀhĀt al-Tafsąr fą MiĆr fąl-ĂAĆr al-Ąadąth*. Cairo: MaтbaĂat al-KalĀną.

Shlomo, Pines. 1986. "What was Original in Arabic Science" in his *Studies in Arabic Versions of Greek Texts and in Mediaeval Science*. Leiden: Brill, pp. 181–205.

Sourdel, D. 1986. "Bayt al-Hikmah." *EI*, 1: 1141.

Steenberghen, Fernand van. 1955. *Aristotle in the West*. Translated by Leonard Johnston. Louvian: Nauwelaerts.

Stenberg, Leif. 1996. *The Islamization of Science: Four Muslim Positions Developing an Islamic Modernity*. Lund: Lund Studies in History of Religion. This book presents a systematic account of modern scholarship on Islam and science through developing a fourfold division of the discourse.

Suhrawardi, Shihab al-Din. Tr. 1999. *Hikmat al-Ishraq*. Translated by John Walbridge and Hossein Ziai as *The Philosophy of Illumination*. Prova: Brigham Young University Press. This is a new critical edition of the text of Suhrawardi's magnus opus, *Hikmat al-Ishraq*. The bi-lingual Arabic-English text is annotated and supplemented by commentary and notes.

as-Suyuti, Jalal al-Din. 1982. *al-hay'a as-saniya fi'l hay'a as-sunniya*. Translated by Anton M. Heinen. Beirut: Franz Steiner Verlag. An important work, this excellent critical edition with translation and commentary by Anton M. Heinen traces as-Suyuti's sources, gives background information on the cosmological tradition in Islam, comments on the nature of scientific enterprise in early centuries of Islam, and provides a translation of as-Suyuti's work along with the original Arabic.

Swartz, Merlin L., ed., 1981. *Studies on Islam*. New York: Oxford University Press.
Taeschner, Fr. 1991. "The Ottoman Geographers." *EI*, 2: 587–590.
Tantawi, Jawhari. 1931. *al-Jawahir fi Tafsir al-Qur'an al-Karim al-Mushtamil al Aja'ib*, 26 vols. Cairo: Mustafa al-Bab al-Halabi.
Toomer, G. J. 1996. *Eastern Wisedome and Learning, The Study of Arabic in Seventeenth-Century England*. Oxford: Clarendon.
Trude, Ehlert. 1993. "Muhammad." *EI*, 7: 360–387.
Vernet, J. 1997. "Al-Khwarazmi." *EI*, 4: 1070.
Watton, William. 1694. *Reflections upon Ancient and Modern Learning*. London: J. Leake. The full title includes the following lines: With a dissertation upon the epistles of Phalaris, Themistocles, Socrates, Euripides; &c. Originally printed when Watton was 28, as a reply to Sir William Temple's *Ancient and Modern Learning* (1692), and as a defense of the kind of scholarship then under attack as pedantry. The primary text deals with various aspects of ancient and modern science, mathematics, philosophy, and philology.
Youschkevitch, A. P. and B. A. Rosenfeld. 1980. "al-Khayami." *DSB*, 7: 323–334.
Ziadat, Adel A. 1986. *Western Science in the Arab World, The Impact of Darwinism*, 1860–1930. London: Macmillan. This is a somewhat dated survey of the arrival of Darwinism in a part of the Arab world. Based on original sources, it provides good insights into the making of the discourse on Darwinism in certain parts of the Arab world.

Index

Abbasids, 11, 12, 15, 18, 24, 25, 36, 37, 73, 77, 113, 126, 131, 217
Abduh, Muhammad, 140, 152, 157
al-Abhari, Athir al-Din, 121
Abu Bakr, 4
Abu Ma'shar, xx, 106
Abu Qurra, Theodore, 25
Abulfeda, 115
Adelard of Bath, 106
Aeneas, 114
afaq, sing. *falq*, 39
al-Afghani, Jamal al-Din, 71, 146–50, 157
Afghanistan, 33, 145, 146
Afghans, 136
Ahmad, Hanafi, 152
Ahmad, Maqul S., 36–38, 40
Aleppo, 116, 214
Algebra, 42–44, 65, 106, 107
Algorithm, 42
Aligarh Muslim University, 144
al-Andalus, xxiii, 20, 36, 109, 110, 112, 113
Almagest, 51, 106, 107
Anees, Munawwar, 170
Apollonius, 44
Appropriation, 25, 27
aql, 196
Arabic 42, 46, 48, 62, 64, 96, 97, 135, 137, 144; language, 153; *lingua franca* 153, 154; manuscripts, 11, 114; science, 127, 149; studies, 114, 115; translation movement, 103–18; translations 15, 16, 19, 22, 25, 31, 34, 36, 39, 40, 56, 68, 76, 83, 93, 103–18, 122, 128, 146, 156–58, 164
Arabists, 126
Architecture, 18, 133, 175
Aristotle, belief in eternity 83; classification scheme 22; commentators of, 77, 108; concept of God, 23; cosmology, 84–86; influence of, 23, 37, 83, 85–93, 190; natural philosophy, 68; physics 53; reception of 83; refutation of 56; transformation of 29, 32, 33, 79, 80; translations of his works 25, 107
Astronomy, 46–53, 107, 108; Copernican system, vii; folk, 36; Greek 16; Islamic, 11; Latin, 109; theoretical foundations of, 10
Astrolabe, 48, 53, 55, 111
Atto, the bishop of Vich, 104

Authority, xvii, 25, 68, 70, 74, 89, 120, 163, 177, 178
Averroes, 72, 75, 76, 77, 205, 213
ayat, sing. *aya*, 6, 30, 41, 90, 91, 156, 163, 176, 199, 201
Azzindani, Abdul Majeed A., 163

Bacon, Francis, x, 116, 214
Baghdad, 11, 12, 15, 25, 33, 38, 39, 41, 50, 78, 96, 98, 124, 131, 135, 189
al-Balkhi, 40
Banu Musa, 107
Barbour, Ian, xv, xvi
al-Battani, 47
Bengal, 133
Big Bang, 164
al-Biruni, 23, 33–36, 56–58, 66, 67, 73, 74, 89–91, 105, 120, 121, 138, 193, 198
al-Bitruji, xx, 50, 108
Bologna, 109
Bombay, 146
Botany, 24, 58, 163, 207
Boyle, Robert, 63
Brahe, Tycho, 52
Bramah, Joseph, 132
Bucaille, Maurice, 164, 165
Bucaillism, 170
Bukhara, 86, 87, 198

Caesar, 114
Cairo, 48, 96, 135, 152, 185
Calcutta, 146
Caliphate, 11, 12, 138
Canon of Medicine, 87, 106, 108, 198, 201
Cantor, Geoffery, xvi
Catching-up Syndrome, 182
Causality, viii, 56, 98, 190, 206
China, 126
Chittick, William C., 95, 97, 98
Christianity, ix, xvi, xvii, 63, 64, 70–72, 75, 83, 104; and science, xvii, 61, 63, 107, 141
Cicero, 114
Colonization, xvii, xviii, 71, 118, 133, 134, 136, 139, 140, 159, 165
Commission for Scientific Miracles of Qur'an and Sunna, 162

Constantine the African, 104
Constantinople, 137
Copernicus, ii, 50, 109
Cosmology, 9–15, 29, 31, 34, 80–91, 95–98, 127, 153, 163; Aristotelian, 30, 82, 84; Peripatetic, 86; Sufi, 95
Cotton, 206
creatio ex nihilo, 79, 82
Creation, al-Biruni's view, 57, 90; Aristotle's view, 22, 85; Christian, ix, 30; creation versus eternity, 91–102; critique of creationism, 170; *ex nihilo*, 32, 89; Mu'tazilah and Asha'irah view, 31; Qur'anic account, 5, 7, 30, 56, 79, 80–82; science, ix; scientific exegeses, 153, 157, 164, 165
Crusades, 2, 112, 113

Damascus, 12, 25, 44, 48, 50, 96
Dante Alighieri, 113
Dar al-Harb, 142
Darwin, Charles, ii, vii, viii, xiii, 68, 146, 153
Darwinism, 146, 153–58
Davidson, Herbert A., 92–94
Delhi, 96, 136, 143
al-Dhahabi, Muhammad Husayn, 216
din, 62, 65, 95
Dutch, 114, 132

East India Company, 133
Eaton, Charles Le Gai, 172
Egypt, 131, 133, 135, 155, 156, 181, 184, 186
Einstein, Albert, 63, 169
Electra, 114
Empedocles, 73
Essence, *mahiya*, 31, 83, 85, 87, 100, 191, 197, 203
Euclid, 44, 87, 107
Evolution, vii, xii, 68, 154–58, 168, 176, 178
Exegesis, xxiv, 96, 151, 152

falasifa, 106, 216
al-Farabi, Abu Nasr, 23, 69, 87, 105, 107
al-Farghani, 37, 107

al-Faruqi, Ismail, 167
fatwa, 75, 141, 182, 184
al-Fazari, 11, 12, 37, 53
Feingold, Mordechai, 114, 116
Fez, 96
fiqh, 76
Fuchs, Leonhart, 118

Galen, 16, 116, 201
Galileo, Galilei, vii, viii, xiii, xvii, 61, 70, 72, 117, 149, 150
Generation and Corruption, 196
Geography, 9, 19, 29, 34–43, 76, 122, 137; Balkhi School, 38; Islamic cartography, 19, 35, 40
Geology, 57, 58, 122, 153, 163
Gerard of Cremona, 106, 108, 198
Gerbert of Aurillac, 104
al-Ghazali, Abu Hamid, 23, 86, 94, 95, 98, 120, 124, 126, 150, 205–11
Gnostic, 88, 98, 101
God, xx, 5–8, 22, 36, 62, 63, 73, 85, 88, 141, 144, 164, 176–78, 202, 204; Aristotle's concept of, 22, 23; attributes of, 7, 31, 51, 81, 83, 84, 90; and causality, 206, 210, 211; as Creator, 7, 30, 79, 82, 85, 91–98, 157, 158, 170; and Illuminationist philosophy, 89; and Muslim scientists, 41, 42, 47, 48, 57, 87, 90, 151, 152; and *Mutikallimun*, 56; and nature, 6–8; *Tawhid* (Oneness of), xviii, 80–82, 86
Goldziher, Ignaz, 20, 64, 70–75, 125–28, 147, 148
Goldziherism, 74, 128
Grant, Edward, 217
Greaves, John, 115
Greek philosophy, xix, 74, 77, 86
Guilds, 11
Guénon, René, 172–74
Gutas, Dimitri, 18, 24, 25, 70, 71, 217

Habash al-Hasib, 47
Hadith, xxiv, 4, 59, 73, 95, 112, 163

Haeckel, Ernst Heinrich, 158
al-Hamawi, Yaqut, 39, 41
Hanafi, 141, 152
Hanbali, 71
Haq, Syed Nomanul, 7, 8, 32
Hassan Hussein, 158
Heisenberg, Werner, 160
Hermeneutics, 24, 92, 165
hikmah, 74–76, 175, 177
Hindu, 11, 42–44, 73, 90
Hodgson, Marshall G. S., 134–36
Hoodbhoy, Pervez, 77
Huff, Toby E., 64, 126, 127
Hugh of Santalla, 105
Human being, 6, 7, 23, 62, 67, 91, 153, 181, 184, 185, 201, 202
Humanists, 115, 116
Hunayn ibn Ishaq, 25, 107
Husayn, bin Ali 11
Hussein al-Jisr, 157

Ibn al-Arabi, 215
Ibn al-Bitriq, 25
Ibn al-Haytham, 16, 17, 23, 47, 50, 111, 138
Ibn al-Muqaffa, 25
Ibn al-Nafis, 121
Ibn al-Shatir, 48, 50, 52, 121, 139
Ibn Battuta, 38
Ibn Jubayr, 38
Ibn Khaldun, 23
Ibn Khurradadhbih, 37, 38
Ibn Na'ima, 25
Ibn Rushd, 23, 32, 50, 69, 73–77, 86, 87, 94–98, 107, 108, 113, 126, 201, 205–11
Ibn Sina, xx, 23, 32, 33, 58, 69, 73, 77, 80, 86–95, 98, 101, 106, 108, 113, 138, 139, 193, 194, 198–201
Ibn Tufayl, 92, 201–5
Ibn Yunus, 48
Ibrahim Pasha, 137
al-Idrisi, 39
Ijmalis, 168, 170
ijtihad, 183, 184
Ikhwan al-Safa, 158, 221
Intellectual Sciences, 100

Index

International Institute of Islamic Thought, 167
Iqbal, Muzaffar, 73, 193
al-Iraqi, Fakhr al-din, 95
Isfahan, 44, 100, 119, 123, 124, 136
al-Isfahani, Abu al-Majid Muhammad Rida, 157, 158
al-Iskandarani Muhammad ibn Ahmad, 151
Islam and Science, xix, 1, 5, 8, 17, 20, 21, 61–66, 70, 72, 77, 78, 101, 103, 122, 125, 129, 131–33, 168, 170, 175, 177, 182, 186, 188, 206; colonial era, 135–40; Islam as justifier of science, 140, 141–51, 162–66; modern approaches, 139, 160–62, 187, 188; nexus, 53, 55, 67, 68, 73, 78–80, 146, 199, 200; scientific *tafsir*, 151–53
Islamic art, 53
Islamic astronomy, 11, 15, 48, 50
Islamic jurisprudence, 185
Islamic orthodoxy, 74, 122, 124, 127
Islamic philosophy, 56, 74, 98, 99, 140, 177
Islamic science, 16, 18, 20, 26, 71, 73, 74, 103, 125, 128, 170
Islamic Scientific Tradition, xvi, xvii, xix, xxiii, 10, 15, 16, 19, 20, 23, 24, 26, 27, 30, 44, 50, 53, 66, 102, 107, 116, 118; decline of, 120–29
Islamization, 78, 167
Islamize, 79, 167
Ismail Mazhar, 158
Istanbul, 51, 52
Izutsu, Toshihiko, 6, 101

Jabir bin Aflah, 50
Jabir ibn Hayyan, 9, 12, 14, 15, 32, 50, 93, 107
al-Jahiz, xxi
Jansen, J. J. G., 152
Jawhari, Tantawi, 152
al-Jazari, 121
Jesuits, 155
John of Seville, 106

al-Juwayni, Abu'l-Ma'ali Abd al-Malik, 93
al-Juzjani, 50

Kalam, 94
Kalin, Ibrahim, 166, 173, 175
Kant, Immanuel, viii, 61
al-Kashi, Jamshid Ghiyath al-Din, 42, 44, 48, 121
Katibi, 98
Kemal, Namik, 150
Kennedy, E. S., 15, 27, 34, 48, 50
Kenny, Chris, xvi
Kepler, Johannes, xvii, 149
Khan, Nadir, 136
Khan, Sayyid Ahmad, 140, 141, 145, 149
Khawarij, 11
al-Khazini, Abd al-Rahman, 56
al-Khwarizmi, 37, 42–44, 65, 67, 107, 109, 139
al-Kindi, Ya'qub ibn Ishaq, 23, 25, 33, 37, 43, 58, 69, 77, 78, 107, 189–93
King, David, 19, 121
Kitab al-Manazir, 17
Kitab al-Tawhid, 158
Knowledge, *ilm*, 21, 23, 46, 64, 67, 73, 106, 133
Kufa, 12, 189
Kuhn, Thomas, 126
Kushyar bin Labban, 43

Latin translations, 13, 19, 21, 41, 42, 69, 75, 87, 103–9, 112–16, 128, 132, 198, 201, 205
Lewis, Bernard, 220
Lindberg, David, 125
Lings, Martin, 172, 176
Logic, 3, 37, 48, 75, 76, 87, 89, 93, 107, 201

Madinah, 2–3, 4, 9, 10, 12, 18, 35
Maimonides, 79, 80, 92
Makdisi, George, 70
Makkah, 1–3, 11, 19, 34, 35, 38, 40, 46, 47, 96, 135, 141, 146

Index

Malik, Hafeez, 220
al-Ma'mun, 25, 37, 40, 50
al-Manar, 152
al-Mansur, 11, 12, 25, 36, 131
Manzoor, Pervez S., 170
Maragha, 50, 52
al-Mashriq, 157
Mathematics, 15, 22, 24, 34, 35, 46, 47, 72, 104, 107, 124, 128, 137, 150, 183
mawaqeet, 19
McVaugh, Michael, 105
Medicine, 14, 15, 46, 58, 59, 73, 76, 87, 104, 105, 107, 108, 118, 121, 137, 163, 183, 198–201, 205, 207, 218; Prophetic, 9, 219
Merv, 12, 36
Metaphysica, 83
Metaphysics, 22, 37, 87, 91, 100, 166, 171, 176, 177, 200
Mineralogy, 58, 207
Mir Damad, 100
Miracle, 210, 211
Missionaries, 155
mithaq, 7
Modern science, xvi, xvii, xviii, xix, xxiii, 1, 31, 68, 97, 121, 127, 128, 135, 139–61, 164–78, 182–89
Modernity, xix, 140, 167, 173, 187, 188
Monastery, 50, 104, 105
Mongol, 95, 98, 124, 131, 134
Moore, Keith, 163, 166
Moral and ethical issues, 68; biogenetics, 170–88; *fatwas*, 184
Mu'tazilah, 31
Muhammad, the Prophet, xxv, 2, 5, 19, 62, 71, 112, 113, 143, 164, 170
Muhammad Ali, 135, 136
Muir, William, 143
Musa, Mark, 215
Musnad of Ahmad, 73
Mustafa Kemal, 137, 138, 150
mutakallimun, 56

Napoleon, 133, 135

Nasr, Seyyed Hossein, 6, 8, 33, 41, 83, 87–90, 98, 100, 101, 172, 175, 176
Nature, innate (*fitrah*), 62, 202; work of God, 141, 144
Necessary Being, 88, 100
Necheris, 145–47
Newton, Isaac, 63, 132, 149
Novum Organum, 116
Nursi, Badiuzzeman Said, 140, 150, 151

Observatories, Maragha, 50, 52; Samarqand, 43, 44, 48, 49, 51, 67
Olson, Richard, xiii, xxiv, 61, 63
Ontology, 166, 169
Optics, 56, 107, 214
Organization of Islamic Conference, 185
Organon, 86
Orientalism, 112–18; orientalists, 15, 18, 127
Orpheus, 114
Orthodoxy, 70–74, 124
Ottoman Empire, 135; Tulip Age, 133
Ottomans, 115, 131–38, 156
Oxford, 109, 115, 144

Pagel, Walter, 222
Pahlavi dynasty, 24, 40, 137
Paris, 71, 109, 114, 147, 159
Pasha, Ibrahim, 137
Peripatetic, 53, 56, 88–98, 194
Perry, Whitall, 176, 177
Philoponus, 83, 93
Pingree, David, 11, 15
Plato, 37, 83, 84, 114
Plato of Tivoli, 106
Plotinus, 32, 33, 83, 84
Pope Sylvester II, 104
Portuguese, 132
Posterior analytics, 107
Precursorism, 26
Principia, 132
Prophet of Islam, 2, 19, 58, 62, 65, 73, 80, 112
Psychology, 37, 58
Ptolemy, 36, 37, 50–52, 87, 106, 107

Qajar, 136
qiblah, 3, 35–37, 40, 46, 47
Quiddity, 100
Qum, 124
Qur'an, 4–8, 10, 19, 21, 31, 35, 36, 39, 46, 57, 58, 62–65, 80, 82, 84, 96, 97, 133, 135, 144, 183; account of the origin, 31, 57, 79; concept of God, 23; concept of knowledge, 22; exegesis, xxiv; Latin paraphrase of, 114; and modern science, 150, 151; Qur'anic cosmology, 15, 81–84, 89; Qur'anic cosmos 30; Qur'anic studies 11; scholars, 15; sciences of, 4, 80; scientific miracles of, 162–64; scientific *tafsir*, 151–53, 158; *tafsir*, 140; translations of, 114; view of nature 5–7, 18, 90; worldview, xviii, 41, 82

Ratzsch, Del, xvi
Rashed, Roshdi, 149
Radiant Cosmography, 81–83
al-Razi, Abu Bakr Zakaria, xxvii, 13, 14, 16, 56, 77, 89, 107
al-Razi, Fakhr al-Din, xxx, 56, 94
Reductionism, 26
Religion and science, 17, 63, 64, 67, 71, 73, 78, 79, 97
Renaissance, 20, 27, 61, 63, 105, 113, 115, 128, 173
Renan, Ernest, 71, 72, 147–50, 159
Revelation, 2, 25, 26, 67, 75, 171, 172, 177, 211
Rida, Rashid, 140, 152, 157
Risale, 150
Robert of Chester, 41, 106
Robert of Ketton, 114
Roman Empire, 105
Royal Asiatic Society of London, 143
ruh, 106
Rumi, Jalal al-Din, 39, 95

Saadia, 215
Sabra, A. I., 16, 26, 27, 70, 121, 122
Sabsvari, Mulla Hadi, 98
Safavi, 119, 131, 134

Safavid, 123, 124, 136
Saliba, George, 10, 15, 16, 50
Salim III, 133
Samarqand, 43, 44, 48, 49, 51, 67
Al-Samawal, 44
Sanskrit, 11, 36
Sardar, Ziauddin, 168–70
Sarton, George, 105, 121, 198
Sayili, Aydin, 47, 50, 129
School of Illumination, 83, 89, 98
Schuon, Frithjof, 172–75
Schweigger, Solomon, 114
Science and religion, 61, 63, 65, 79, 101, 108
Science of nature, 174
Scientific journals, 155, 156
Scientific Revolution, 1, 64, 68, 120, 125–27, 149, 172
Scientism, 188
Shah Abbas, 123
Shah Wali Allah, 134
al-Shirazi, Qutb al-Din, 50, 52, 95, 98, 121
Shirazi, Sadr al-Din, 98, 99
shukuk literature, 16
Signs, *ayat*, 6, 30, 41, 90, 91, 156, 163, 176, 199, 201
Sirhindi, Shaykh Ahmad, 96
Sophia Perennis, 172
Spain, 12, 19, 20, 36, 87, 95, 98, 104–10, 112, 137
Starkey, George, 116
Stenberg, Leif, 166
Sufis, 29, 31, 82, 89, 95, 96
Sunna, 184
as-Suyuti, 82, 224
Swartz, Merlin L., 216
Syria, 2, 131, 155, 161
Syrian Protestant College, 155

Tabataba'i, Sayyid Muhammad Husayn, 140
tafsir, 140, 151, 152
tafsir al-ilmi, 140, 151
Tawhid, 8, 62, 80, 86, 147, 171
Teleology, 90, 172
Thabit ibn Qurrah, 25, 43

Theology, 74, 100
Thomas Aquinas, 79, 92
Timaeus, 37
Timur, 44
Tipu Sultan, 133
Toledo, 105–7, 109, 111
Traditionalists, 171–78
Translation Movement, 15; first phase, 104–8; second phase, 109–12; third phase 112, 114–18
Transoxania, 131
Tulip Age, 133
Turkey, 131, 137, 138, 140, 150, 158, 186
al-Tusi, Nasir al-Din, 50, 52, 95, 121
Two-entity model, 65, 68

Ulugh Beg, 44, 48, 49, 51, 54
Umayyads, 12

al-Uqlidisi, 44
Uthman b. Affan, 11, 25, 134
Uthman Dan Fodio, 134
Wajdi, Muhammad Farid, 152
Wali Allah, Shah, 96
Watt, James, 132
Watton, William, 116
Weber, Max, 126
William of Moerbeke, 108
World Muslim League, 162
wujud, 96

zakah, 3, 44
Zand, Karim Khan, 136
Ziadat, Adel A., 157
zij, 11, 44, 47, 106
Zoology, 24, 58, 207
Zoroastrians, 10

About the Author

MUZAFFAR IQBAL is the president of the Center for Islam and Science (www.cis-ca.org), Canada. A scientist by training, an Islamic scholar by vocation, a novelist and a poet, his previous works include *Islam and Science* (2002) and *God, Life, and the Cosmos: Christian and Islamic Perspectives* (coeditor, 2002).